9/24

Science, Technology and Medicine in Colonial India

Interest in the science, technology and medicine of India under British rule has grown in recent years and has played an ever-increasing part in the reinterpretation of modern South Asian history. Spanning the period from the establishment of East India Company rule through to Independence, David Arnold's wide-ranging and analytical survey demonstrates the importance of examining the role of science, technology and medicine in conjunction with the development of the British engagement in India and in the formation of Indian responses to Western intervention. One of the first works to analyse the colonial era as a whole from the perspective of science, the book investigates the relationship between Indian and Western science, the nature of science, technology and medicine under the Company, the creation of state scientific services, 'imperial science' and the rise of an Indian scientific community, the impact of scientific and medical research and the dilemmas of nationalist science.

DAVID ARNOLD is Professor of South Asian History at the School of Oriental and African Studies, London. His publications include *The Problem of Nature* (1996) and *Colonizing the Body: State Medicine and Epidemic Disease in Nineteenth-Century India* (1993).

THE NEW CAMBRIDGE HISTORY OF INDIA

General editor GORDON JOHNSON

President of Wolfson College, and Director, Centre of South Asian Studies,
University of Cambridge

Associate editors C. A. BAYLY

Vere Harmsworth Professor of Imperial and Naval History, University of Cambridge,
and Fellow of St Catharine's College

and JOHN F. RICHARDS

Professor of History, Duke University

Although the original **Cambridge History of India**, published between 1922 and 1937, did much to formulate a chronology for Indian history and describe the administrative structures of government in India, it has inevitably been overtaken by the mass of new research over the last sixty years.

Designed to take full account of recent scholarship and changing conceptions of South Asia's historical development, **The New Cambridge History of India** is published as a series of short, self-contained volumes, each dealing with a separate theme and written by one or two authors. Within an overall four-part structure, thirty complementary volumes in uniform format will be published. Each will conclude with a substantial bibliographical essay designed to lead non-specialists further into the literature.

The four parts planned are as follows:

I The Mughals and their contemporaries

II Indian states and the transition to colonialism

III The Indian Empire and the beginnings of modern society

IV The evolution of contemporary South Asia

A list of individual titles in preparation will
be found at the end of the volume.

THE NEW CAMBRIDGE HISTORY OF INDIA

III · 5

Science, Technology and Medicine in Colonial India

DAVID ARNOLD

School of Oriental and African Studies
University of London

PUBLISHED BY THE PRESS SYNDICATE OF THE UNIVERSITY OF CAMBRIDGE
The Pitt Building, Trumpington Street, Cambridge, United Kingdom

CAMBRIDGE UNIVERSITY PRESS
The Edinburgh Building, Cambridge CB2 2RU, UK   www.cup.cam.ac.uk
40 West 20th Street, New York, NY 10011–4211, USA   www.cup.org
10 Stamford Road, Oakleigh, Melbourne 3166, Australia
Ruiz de Alarcón 13, 28014 Madrid, Spain

© Cambridge University Press 2000

This book is in copyright. Subject to statutory exception and to the provisions of relevant collective licensing agreements, no reproduction of any part may take place without the written permission of Cambridge University Press.

First published 2000

Printed in the United Kingdom at the University Press, Cambridge

Typeface Garamond MT 10.5/13pt *System* QuarkXpress™ [SE]

A catalogue record for this book is available from the British Library

Library of Congress Cataloguing in Publication data

Arnold, David, 1946–
Science, technology, and medicine in India, 1760–1947 / David Arnold.
p. cm. – (New Cambridge history of India ; III, 5)
Includes bibliographical references and index.
ISBN 0 521 56319 4 (hb)
1. Science – India – History.   2. Technology – India – History.
3. Medicine – India – History.   I. Title.   II. Series.
DS436.N47   pt. 3, vol. 5   [Q127.I4]   509.54—dc21   99-27791   CIP

ISBN 0 521 56319 4 hardback

CONTENTS

List of illustrations	*page* vi
List of tables	vii
General editor's preface	viii
Preface	x
List of abbreviations	xii

1	Introduction: science, colonialism and modernity	1
2	Science under the Company	19
3	Western medicine in an Indian environment	57
4	Technologies of the steam age	92
5	Imperial science and the Indian scientific community	129
6	Science, state and nation	169
	Conclusion	211

Biographical notes	214
Bibliographical essay	217
Index	227

ILLUSTRATION

1 Plan of the H.E.I.C.'s Botanic Garden, Saharunpore, 1831 (reproduced by kind permission of the Wellcome Trust)

page 54

TABLES

5.1	Composition and pay of India's scientific services, 1920	*page* 138
6.1	The founding of science departments at selected Indian universities	191
6.2	Indian Fellows of the Royal Society, 1918–47	194

GENERAL EDITOR'S PREFACE

The New Cambridge History of India covers the period from the beginning of the sixteenth century. In some respects it marks a radical change in the style of Cambridge Histories, but in others the editors feel that they are working firmly within an established academic tradition.

During the summer of 1896, F. W. Maitland and Lord Acton between them evolved the idea for a comprehensive modern history. By the end of the year the Syndics of the University Press had committed themselves to the *Cambridge Modern History*, and Lord Acton had been put in charge of it. It was hoped that publication would begin in 1899 and be completed by 1904, but the first volume in fact came out in 1902 and the last in 1910, with additional volumes of tables and maps in 1911 and 1912.

The *History* was a great success, and it was followed by a whole series of distinctive Cambridge Histories covering English Literature, the Ancient World, India, British Foreign Policy, Economic History, Medieval History, the British Empire, Africa, China and Latin America; and even now other new series are being prepared. Indeed, the various Histories have given the Press notable strength in the publication of general reference books in the arts and social sciences.

What has made the Cambridge Histories so distinctive is that they have never been simply dictionaries or encyclopaedias. The Histories have, in H. A. L. Fisher's words, always been 'written by an army of specialists concentrating the latest results of special study'. Yet as Acton agreed with the Syndics in 1896, they have not been mere compilations of existing material but original works. Undoubtedly many of the Histories are uneven in quality, some have become out of date very rapidly, but their virtue has been that they have consistently done more than simply record an existing state of knowledge: they have tended to focus interest on research and they have provided a massive stimulus to further work. This has made their publication doubly worthwhile and has distinguished them intellectually from other sorts of reference book. The editors of *The New Cambridge History of India* have acknowledged this in their work.

The original *Cambridge History of India* was published between 1922 and 1937. It was planned in six volumes, but of these, volume 2 dealing with the period between the first century AD and the Muslim invasion of India never appeared. Some of the material is still of value, but in many respects it is now out of date. The past fifty years have seen a great deal of new research on India, and a striking feature of recent work has been to cast doubt on the validity of the quite arbitrary chronological and categorical way in which Indian history has been conventionally divided.

The editors decided that it would not be academically desirable to prepare a new *History of India* using the traditional format. The selective nature of research on Indian history over the past half-century would doom such a project from the start and the whole of Indian history could not be covered in an even or comprehensive manner. They concluded that the best scheme would be to have a *History* divided into four overlapping chronological volumes, each containing short books on individual themes or subjects. Although in extent the work will therefore be equivalent to a dozen massive tomes of the traditional sort, in form *The New Cambridge History of India* will appear as a shelf full of separate but complementary parts. Accordingly, the main divisions are between I. *The Mughals and Their Contemporaries*, II. *Indian States and the Transition to Colonialism*, III. *The Indian Empire and the Beginnings of Modern Society*, and IV. *The Evolution of Contemporary South Asia*.

Just as the books within these volumes are complementary so too do they intersect with each other, both thematically and chronologically. As the books appear they are intended to give a view of the subject as it now stands and to act as a stimulus to further research. We do not expect the *New Cambridge History of India* to be the last word on the subject but an essential voice in the continuing discussion about it.

PREFACE

Since the first *Cambridge History of India* appeared several decades ago, and more especially over the last ten to fifteen years, there has been a rapid growth of scholarly interest in the history of science, technology and medicine in modern South Asia. This interest has arisen from diverse sources and agendas – from a lateral extension of the history of science from Europe into colonial waters; from an attempt to reassess the impact of the West on India (or conversely to demonstrate the survival and adaptation of pre-colonial ideas and practices) and to explore the nature and consequences of India's modernity; from a new willingness to contest the supposedly objective (or triumphalist) claims previously made in the name of science, technology and medicine; from a desire to probe the origins (and disturbing consequences) of an alliance between science, technology and the state in contemporary South Asia; or simply from a growing recognition of the extraordinary range and richness of the sources available to students of South Asia. As with any relatively new field of enquiry, certain issues have commanded extended, in-depth discussion, while others, seemingly of no less significance, have been largely ignored. In general, medicine and public health have been extensively and critically reappraised, whereas many aspects of science, and more especially technology, remain remarkably unexplored except in ageing works of the high colonial era. In general, we know more about state policies than we do about public responses, a great deal about the official face of science, technology and medicine, but relatively little about how it impacted on individual lives and specific social strata.

This book attempts to provide a broad interpretative overview of the scientific, technological and medical developments of the period from the mid-eighteenth to the mid-twentieth centuries, but it makes no claim to be comprehensive. It offers a thematic approach and a broad periodisation, centring on the watershed years of the 1890s and 1900s. In common with much of the recent historiography from which it draws, it seeks to examine the social and ideological context rather than the purely technical nature and significance of the developments concerned, to see science as a cultural construction, with political agendas and institutional structures of its own, and not merely as an objective phenomenon or a subject for individual biography. It aims to make central to the discussion the complex interrelationship between India and the

PREFACE

West, rather than speaking (as if unproblematically) of technology 'transfers' or the 'spread' of Western scientific ideas and medical practices. It seeks to identify the diverse elements of science, technology and medicine, drawn from different traditions, that informed ideas of India's past and modernity.

In researching this book over several years I have been greatly assisted by invaluable advice (and offprints) from, among many others, Michael Adas, Robert Anderson, Neeladri Bhattacharya, Partha Chatterjee, Indira Chowdhury, Andrew Grout, Richard Grove, Ramachandra Guha, Sumit Guha, David Hardiman, Mark Harrison, Douglas Haynes, Shruti Kapila, Sudipta Kaviraj, Ian Kerr, Deepak Kumar, Claudia Liebeskind, Jim Masselos, John Pickstone, Mridula Ramanna, Peter Robb, Satpal Sangwan, Sumit Sarkar, Sanjay Sharma, Elizabeth Whitcombe and Michael Worboys. In many cases, their work has been the inspiration for what I have attempted to put together here and I hope I have done justice to the originality of their research and theorising. I am also greatly indebted to the Wellcome Trust and the Leverhulme Trust for funds to support sabbatical research and to visit India in 1997, and to archives and libraries in Britain and India, especially the now defunct India Office Library, the libraries of the School of Oriental and African Studies and the University of London, the National Archives of India and the Nehru Memorial Museum and Library in Delhi, the Tamil Nadu Archives in Madras, the National Library in Calcutta, and the Rockefeller Archive Center, Tarrytown, New York. My sincere thanks, too, to Chris Bayly and Gordon Johnson for agreeing to include science, technology and medicine in the *New Cambridge History of India* and to Marigold Acland at Cambridge University Press for her encouragement and patience. Finally, thanks once again to Juliet Miller – for help with nature and the Nobel prizes, and for the pleasures and diversions of our life together.

ABBREVIATIONS

AR	*Asiatic Researches*
BHM	*Bulletin of the History of Medicine*
BJHS	*British Journal for the History of Science*
CJNH	*Calcutta Journal of Natural History*
CR	*Calcutta Review*
ICS	Indian Civil Service
IESHR	*Indian Economic and Social History Review*
IJHS	*Indian Journal of History of Science*
IMG	*Indian Medical Gazette*
IMS	Indian Medical Service
IRFA	Indian Research Fund Association
JAS	*Journal of Asian Studies*
JASB	*Journal of the Asiatic Society of Bengal*
JRAS	*Journal of the Royal Asiatic Society*
MAS	*Modern Asian Studies*
MH	*Medical History*
MR	*Modern Review*
NAI	National Archives of India (New Delhi)
OIOC	Oriental and India Office Collections (British Library, London)
RAC	Rockefeller Archive Center (Tarrytown, New York)
SAR	*South Asia Research*
SH	*Studies in History*
TNA	Tamil Nadu Archives (Madras)

CHAPTER 1

INTRODUCTION: SCIENCE, COLONIALISM AND MODERNITY

The questions that can be asked about science in modern India are essentially those pertaining to the history and sociology of science elsewhere. What is the social character of scientific knowledge? Who produces science and why? How does science exercise authority within a society and across cultural divides? As historians and sociologists have begun to investigate science, less in terms of its self-declared aims and putatively objective interrogation of nature and more in terms of its internal ordering, social construction and cultural authority, it has become clear that science is 'a highly social activity', one that cannot be 'sealed off from the values of the society in which it is practised'.[1] It is increasingly recognised, too, if not yet universally accepted, that science, far from being monolithic, manifests itself across time and cultures in myriad forms, reflecting as much as informing a given society's cultural, economic and political modalities. Science thus 'reveals itself as much more contingent and culturally specific' than it was once assumed to be.[2] Individuals and groups produce scientific knowledge not in isolation but 'against the background of their culture's inherited knowledge [and] their collectively situated purposes' as well as through 'the information they receive from natural reality'.[3]

The social character and cultural plurality of science has a particular bearing on the history of science, technology and medicine in India, which had a well-established scientific and technological tradition of its own long before being subjected to an extended period of European colonial rule. Although the history of science, technology and medicine continues to be presented in general histories as a record of Western discovery and dissemination, it has become more widely acknowledged than a generation or two ago that not all such histories can be conflated into a single story of European achievement or saga of European enterprise overseas. Particular attention has been directed to understanding the place of science in the colonial world of the eighteenth, nineteenth and early twentieth centuries, in situations in which the history of

---

[1] Nancy Leys Stepan, *'The Hour of Eugenics': Race, Gender, and Nation in Latin America* (Ithaca, 1991), p. 9; Michael Mulkay, 'Sociology of Science in the West', *Current Sociology*, 23 (1980), pp. 1–184.
[2] Stepan, *'Eugenics'*, p. 10.
[3] Steven Shapin, 'History of Science and Its Sociological Reconstructions', *History of Science*, 20 (1982), p. 196.

science often appears inseparable from the history of imperialism itself.[4] Yet, at the same time, in order to understand the social authority and cultural context of science, it is necessary to look beyond the imperial system, beyond its ideologies and instrumentalities, and to look at the 'recipient' society and the manner in which Western science was received and situated in relation to indigenous epistemologies and practices. A history of science in India must also be a history of India, not merely a history of the projection of Western science onto India. One of the principal rationales for a work such as this, which seeks to give an interpretative overview of science, technology and medicine in India from the late eighteenth to the mid-twentieth centuries, must be that there is a new recognition of the centrality of science to an effective understanding of the history of India during the period marked by the rise, ascendancy and retreat of British colonialism in South Asia.

It will be argued here, by way of introduction, that there were three main elements that broadly typified science, technology and medicine in India over this 200-year period. Firstly, there were the traditions of India's own science, technology and medicine, themselves subject to wide internal variations and different historical influences and cultural practices, and the legacies these provided for the subsequent era of British rule. Secondly, there was the nature of Western (or 'colonial') science, technology and medicine as practised in India, their social and intellectual impact, their organisational forms and dual relationship to the colonial regime in India and to metropolitan science in Europe. And thirdly, there was the authority of science, technology and medicine as central attributes of India's modernity, drawing upon indigenous as well as Western sources and finding contested expression in both imperial ideology and nationalist agendas. We will briefly consider each of these in turn.

INDIA'S SCIENTIFIC TRADITIONS

It would be erroneous to think of India as having a single scientific tradition. Over the millennia, India became heir to a wide variety of different oral and textual traditions, drawing upon exogenous contacts as well as indigenous roots. This plurality makes it difficult not only to characterise Indian science as a whole but also to determine the precise nature of its interaction with the forms of science and technology emanating from the West by the late eighteenth and early nineteenth centuries. Even within what is often thought of as the 'Hindu' tradition, there were several strands of scientific ideas and

---

[4] Paolo Palladino and Michael Worboys, 'Science and Imperialism', *Isis*, 84 (1993), p. 102.

practices, including a tradition of empirical, observational science (particularly developed in astronomy and medicine) that functioned alongside, and often in tandem with, various cosmological and astrological beliefs. Whereas astronomy in Vedic India was often closely connected with religious practice (because an accurate knowledge of equinoxes and solstices was needed for the proper timing of sacrifices and other rites), in the post-Vedic and early medieval period the study of astronomy, trigonometry and algebra saw a partial move away from the earlier stimulus of religion and ritual. Thus, one of the most important texts of the later period, the *Surya Siddhanta*, composed around AD 400, devoted a series of chapters to the motion and position of the planets, the nature and timing of eclipses, the rising and setting of the sun and moon, and astronomical instruments such as the armillary sphere; but it also dealt with cosmogony and 'certain malignant aspects of the sun and moon'.[5]

Although the richness and diversity of India's ancient scientific traditions has long been recognised, over the past two centuries it has been the convention to see this as a history of precocious early achievement followed by subsequent decline and degeneration. The European Orientalist scholarship of the late eighteenth and early nineteenth centuries represented India as having had an ancient civilisation equalling, in some respects excelling or anticipating, those of classical Greece and Rome. 'The Asiaticks had climbed the heights of science before the Greeks had learned their alphabet', one enthusiast declared.[6] In astronomy, mathematics and medicine in particular, Hindu science was considered to have been remarkably advanced well before the dawn of the Christian era and to have been the source of discoveries and techniques that were only later taken up and incorporated into Western civilisation, such as 'Arabic' numerals and the use of zero.[7] However, according to this Orientalist interpretation, Indian civilisation was unable to sustain its early achievements and lapsed into decline. There followed an uncritical reliance upon earlier texts: tradition replaced observation as surely as religion supplanted science. This was in part attributed to an increasing rigidity in Hindu society of caste practices and religious belief, but also to the rise of Muslim power in South Asia after AD 1100. Although introducing some scientific and technical skills of its own, Islam was largely seen to have been destructive of the remnants of the old Indian civilisation. The breakup of the Mughal Empire after 1707, the division of India into warring factions and regional

---

[5] For a useful introduction, see Zaheer Baber, *The Science of Empire: Scientific Knowledge, Civilization, and Colonial Rule in India* (New York, 1996), ch. 2.

[6] David Kopf, *British Orientalism and the Bengal Renaissance: The Dynamics of Indian Modernization, 1773–1835* (Berkeley, CA, 1969), p. 102.

[7] Mountstuart Elphinstone, *History of India* (9th edition, London, 1905), pp. 138–60.

states, and the resulting age of 'anarchy' were adduced as further evidence for the stagnation and decay of Indian science, technology and medicine. The history of Indian science thus served as a mere prologue to the eventual unfolding of Western science in South Asia as science was rescued from centuries of decline and obscurity by the advent of British rule and the introduction of the more developed scientific and technical knowledge of the West. This Orientalist triptych – contrasting the achievements of ancient Hindu civilisation with the destruction and stagnation of the Muslim Middle Ages and the enlightened rule and scientific progress of the colonial modern age – has had a remarkably tenacious hold over thinking about the science of the subcontinent. It was a schema deployed not only by British scholars, officials and polemicists but also by many Indians, for whom it formed the basis for their own understanding of the past and the place of science in Indian tradition and modernity. It is still not uncommon for Indian writers to remark, with evident regret, that the 'creative spirit' of Indian science sunk to its 'lowest ebb' between the twelfth and the mid-nineteenth centuries.[8]

Of late, though, some historians of science have sought to break the Orientalist mould. One of the ways in which they have done so has been by looking afresh at the science, technology and medicine of medieval and early modern India, thus revealing the neglected importance of the Muslim contribution to India's scientific traditions or illuminating the emergence, through science, of a dynamic and syncretic Indo-Muslim culture. Medicine has been particularly prominent in this historiographical trend. The mutual enrichment brought about by a 'creative synthesis' between Hindu Ayurveda and Unani-tibb, with its Graeco-Arabic origins, and the apparent absence of rivalry or enmity between its practitioners, the *vaids* and *hakims*, have served to exemplify the continuing vitality and fruitful intermingling of scientific traditions in India well into the eighteenth century, though one might equally argue that Unani physiology and pharmacology were interacting as much with the Indian environment and the region's rich materia medica as with the Ayurvedic system as such.[9] There has also been a new effort to demonstrate that India, far from existing in cultural and technological isolation and being averse to all innovation, had over the centuries borrowed extensively from, and contributed generously to, the scientific and technical knowledge of neighbouring regions, from the Middle East and Central Asia to China and Southeast Asia, and in

---

[8] B. V. Subbarayappa, 'Western Science in India up to the End of the Nineteenth Century', in D. M. Bose, S. N. Sen and B. V. Subbarayappa (eds.), *A Concise History of Science in India* (New Delhi, 1971), p. 484.

[9] R. L. Verma, 'The Growth of Greco-Arabian Medicine in Medieval India', *IJHS*, 5 (1970), pp. 347–63; Baber, *Science*, pp. 78–9.

INTRODUCTION: SCIENCE, COLONIALISM AND MODERNITY

fields as diverse as agriculture, architecture, astronomy, chemistry, medicine, metallurgy, textile production, shipbuilding and armaments.[10] This celebration of cultural eclecticism and trans-regional exchange aligns the history of Indian science more closely with the models of creativity, diffusion and interaction advanced for China, the Muslim Middle East and other non-European culture areas in recent decades, particularly through Joseph Needham's magisterial account of *Science and Civilisation in China* and through other revisionist histories, such as Lynn White's, that have authoritatively established Europe's long-standing debt to Asian technology.[11]

It follows from this revisionist argument that Europe did not impact upon a stagnant and unchanging India. From the late fifteenth century onwards, scientific, medical and technological exchanges continued through the agency and impetus of trade and warfare and through the migration of scholars, merchants, physicians and craftsmen. Contacts flourished in two main directions – with the wider world of Islam (linking India with Iran, Central Asia and the Middle East) but also, increasingly, with the expanding commercial and technological power of Europe. Astronomy, medicine, textiles and arms-making benefited from the fashioning of an Indo-Muslim polity and culture under the Mughals, but India also profited in such areas as shipbuilding and horticulture from contacts after 1498 with the Portuguese and later with the Dutch, French and English.[12] If there remained a gulf between the craft technology of the uneducated artisan and the science of the literati, if there were few individuals before 1750 to whom one could convincingly apply the term 'scientist', then India was in these respects little different from early modern societies in Europe, China or elsewhere. The intellectual activity of religious and cultural elites and the skills of artisans jointly fashioned for India a distinctive place in the annals of science and technology, even if they existed largely in isolation from one another – except when, as for instance at the court of Akbar, the needs of warfare and the prompting of intellectual curiosity brought them temporarily together.[13]

Although from the early sixteenth century the Mughal court was a vital source of patronage for science and technology, dynastic decline in the eighteenth century did not entirely plunge India into obscurity, even if the number

---

[10] S. N. Sen, 'Influence of Indian Science on Other Culture Areas', *IJHS*, 5 (1970), pp. 332–46.
[11] Joseph Needham, *The Shorter Science and Civilisation in China* (ed. Colin A. Ronan), (4 vols., Cambridge, 1978–94); Lynn White, *Medieval Technology and Social Change* (Oxford, 1962).
[12] A. Jan Qaisar, *The Indian Response to European Technology and Culture (AD 1498–1707)* (Delhi, 1982); Irfan Habib, 'The Technology and Economy of Mughal India', *IESHR*, 27 (1980), pp. 1–34.
[13] Irfan Habib, 'Akbar and Technology', in Irfan Habib (ed.), *Akbar and His India* (Delhi, 1997), pp. 129–48.

of manuscripts produced in Sanskrit, Persian and Arabic on scientific and technical subjects showed signs of slowing down.[14] A positive interest in science (and, increasingly, in reconciling the sciences of East and West) flourished under royal patronage in the regional courts of India, from the astronomical observatories built by Raja Jai Singh between 1722 and 1739 at Jaipur, Delhi, Mathura, Ujjain and Benares, to the eclectic medical interests and library of Indian and Western medical texts assembled by Serfoji, the last Maratha ruler of Tanjore.[15] New centres of learning sprang up, some, like Hyderabad under its Nizams or Lucknow under the Nawabs of Awadh, specialising in Islamic science and Unani medicine, while other older, mainly Hindu, seats of learning such as Benares and Nadia in Bengal continued to flourish. Despite the withering away of Mughal power, Delhi remained a significant locus for science, art and literature, and, until the cataclysmic events of 1857, enjoyed a twilight 'renaissance'.[16] There were, however, some areas in which India appeared unresponsive to new technologies. Despite the introduction of the printing press by the Jesuits in Goa in the mid-sixteenth century, it had little influence on India before the late eighteenth century, though its spectacular take-off in the nineteenth century belies any suggestion that this was a consequence of some intrinsic 'mechanical backwardness'.[17] Rather than providing proof of any sustained resistance to technological change, the slowness to adopt printing might rather be taken to indicate the selective manner in which Western science, technology and medicine were appropriated and the persistence of prestigious cultural values, embedded, in this instance, in the manuscript tradition and the skills of artisans and scribes.[18] Matters affecting proficiency in warfare were, by contrast, of more urgent concern and attracted a far more active response. This was the case not only with the Mughals, but also subsequently with the armies of Tipu Sultan of Mysore (until his defeat at Seringapatnam in 1799) and those of Ranjit Singh in Punjab, whose foun-

---

[14] A. Rahman (ed.), *Science and Technology in Medieval India: A Bibliography of Source Materials in Sanskrit, Arabic and Persian* (New Delhi, 1982), pp. xi–xvi.

[15] Baber, *Science*, pp. 85–90; Deepak Kumar, 'Unequal Contenders, Uneven Grounds: Medical Encounters in British India, 1820–1920', in Andrew Cunningham and Bridie Andrews (eds.), *Western Medicine as Contested Knowledge* (Manchester, 1997), pp. 168–82; N. Gangadharam, 'The State of Ayurveda in the Eighteenth and Nineteenth Centuries', in G. Kuppuram and K. Kumudamani (eds.), *History of Science and Technology in India*, 1 (Delhi, 1990), pp. 13–29.

[16] Gail Minault, 'Sayyid Ahmad Dehlavi and the "Delhi Renaissance"', in R. E. Frykenberg (ed.), *Delhi Through the Ages* (Delhi, 1986), pp. 174–85.

[17] L. S. S. O'Malley, 'Mechanism and Transport', in L. S. S. O'Malley (ed.), *Modern India and the West* (London, 1941), p. 221.

[18] For printing and its uses, see C. A. Bayly, *Empire and Information: Intelligence Gathering and Social Communication in India, 1780–1870* (Cambridge, 1996), pp. 235–43.

dries at Lahore and Amritsar manufactured heavy guns and mortars in the 1820s and early 1830s.[19]

Just as it is necessary to rethink the chronology of Indian science and break down the old periodicity of the Orientalist model, so is it imperative to reassess the significance for science, technology and medicine of India's vast land area and internal diversity. Although it is customary and convenient to speak of 'Indian' science or 'Hindu' medicine, such broad aggregations obscure the wide variations between one part of the subcontinent and another. As the examples in the previous paragraph suggest, the decentred nature of India's political and cultural system enabled, most obviously (though not uniquely) in the eighteenth century, several centres of science, technology and medicine to flourish at the same time and for each to develop its own distinctive characteristics. Diversity brought strengths as well as weaknesses. The decline of one centre did not preclude the survival and adaptation of another; India as a whole could profit from the varied intellectual and material products of its different regions and from their interaction and exchange. There were regional schools of Ayurvedic and Unani medicine, just as there were regional variations in the weaving and dyeing of cloth. The physical diversity of the Indian environment, South Asia's almost continental proportions, and the multiplicity of its cultural and political constituencies not only contributed to internal variety and local specialisation but also, from an opposing perspective, challenged attempts (as by British rulers and nationalist scientists) to use the ideological agency and material instrumentality of science, technology and medicine to try to conquer and integrate India's vast interior spaces.

Recurrent, too, in the history of science in India was a tension between the countryside and centres of courtly or regional power, or between cities old and new. Although colonial science might crudely serve to underline the cultural, commercial and political importance of the rise of the three coastal metropolises — Calcutta, Bombay, Madras — this would be to overlook the contribution made to their evolution by the artisans and intellectuals who flocked to them from older centres of manufacturing and scholarship. It would also be to ignore the resilience of other, more ancient centres of learning such as Benares and Delhi. It is not without significance that a number of universities with leading science departments by the 1940s — Lahore, Lucknow, Allahabad, and Dacca, to identify but four points along the Indo-Gangetic axis — were located in cities already prominent on the cultural and political map of India two centuries earlier.

---

[19] Arnold Pacey, *Technology in World Civilization: A Thousand-Year History* (Oxford, 1990), p. 144.

As with cities, so with social groups. Some of India's old intellectual elites resurfaced as agents and interpreters of the new scientific order, as in the case of the Brahmins, Vaidyas and Kayasthas who composed the *bhadralok* (middle-class intelligentsia) in colonial Bengal. It is suggestive, too, of the strength of these intellectual and social continuities that the only Nobel prize to be awarded to an Indian scientist before Independence went to a Tamil Brahmin, C. V. Raman, in 1930. But it should not be overlooked that other social groups (including Parsis, Indian Christians and lower-caste Hindus) also found a place among the practitioners of scientific modernity. The extent to which members of the old intelligentsia brought to their 'modern' avocations skills, insights and inspiration derived from 'traditional' backgrounds (rather than simply trading in their intellectual inheritance to acquire new Western knowledge) is an intriguing issue but one that historians have, as yet, scarcely begun to investigate.[20] Equally, although the advance of British power in South Asia in the late eighteenth and early nineteenth centuries resulted in the overthrow or eclipse of a number of Indian states, culminating in the annexation of Awadh in 1856 and the extinction of Mughal Delhi two years later, it is striking how important India's surviving princes and landed aristocracy were to the patronage of science (in its indigenous as well as Western forms) in the nineteenth and early twentieth centuries and in fields as diverse as astronomy, medicine and technical education. That India's first major hydro-electric scheme was constructed in the princely state of Mysore in 1898 and that ten years later Bangalore became the site for the Indian Institute of Science, should alert us to the significance of even the circumscribed power of the princes in providing an alternative (often more adventurous) source of scientific support and technological initiative to that offered by the British. But if in this respect India's continuing disunity appeared to favour the enterprise of science, in many other respects science in late-colonial India was plagued by the difficulty of trying to create and sustain organisations and institutions that would integrate India into a single scientific entity.

The reappraisal of the character of Indian science, technology and medicine before British rule, therefore, not only is of importance in itself, in establishing the vitality and diversity of an 'indigenous' tradition, but also has wide-ranging implications for understanding what happened after the establishment of the colonial regime. It becomes more difficult to treat India as a kind of scientific and technological *tabula rasa*, whose achievements lay in the

---

[20] Cf. Kapil Raj, 'Knowledge, Power and Modern Science: The Brahmins Strike Back', in Deepak Kumar (ed.), *Science and Empire: Essays in Indian Context (1700–1947)* (Delhi, 1991), pp. 115–25.

remote past and so were unable to affect or inform the course of Western science in South Asia. A recognition of the relative openness and adaptability of India's pre-colonial scientific and technological tradition supports the view that an interactive model might be more appropriate for the colonial period rather than one that depicts either outright confrontation between two intransigent forces or an automatic unassailable Western ascendancy. But, at the same time, pre-colonial science and its legacies should not be asked to explain too much. It is necessary to attach no less importance to the profound rupture caused to Indian society, materially and intellectually, by colonial intervention and the unprecedented impact made by the science, technology and medicine of the West.

COLONIAL SCIENCE

The history of science, technology and medicine in British India has often in the past been represented as essentially the story of the introduction and dissemination of Western ideas, practices and techniques. Such accounts make scant reference either to indigenous scientific, technological and medical traditions (except negatively, as a source of unreasoning and atavistic opposition to the legitimate progress of science, or as a lineage happily long extinct by the late eighteenth century), or to tensions and divergences between science as practised in the colony and that propagated in the capitals of Europe. Of late, however, as the history of science, technology and medicine in India has expanded and as the nature of Western science itself has been subjected to more critical appraisal, the relationship between India and Western science has come to be seen as more complex and less one-directional than previously assumed. The idea of a simple diffusion of a monolithic and progressive Western science into passively recipient extra-European lands has been challenged from several standpoints, not least by a more interactive and regionally focused understanding of how science developed in India from the late eighteenth century onwards.

But it is as well to begin with an ageing orthodoxy. The most influential statement of the diffusionist model of Western science was made by George Basalla in 1967,[21] and though it now appears dated and simplistic in many respects it is still worth summarising as the basis for much of the ongoing discussion of colonial science. How, Basalla asked, did 'modern science' come to be diffused from its original home in Western Europe and 'find its place in the

---

[21] George Basalla, 'The Spread of Western Science', *Science*, no. 156 (1967), pp. 611–22.

rest of the world'? He argued that the process could best be understood through a three-stage model. In Phase One, Europeans established contact with new lands as part of the process of Western reconnaissance, trade, conquest, and colonisation. The 'non-scientific' society served Europe as a source of scientific data, garnered by Europeans through maps and surveys, and mineral, plant and animal specimens. In keeping with Europe's interest at this stage in 'the systematic exploitation of nature', the dominant sciences of Phase One were botany and zoology, followed by astronomy, geology and geography. Although commercial motives provided some impetus for this scientific reconnaissance, Basalla attached more significance to the scientific culture from which Europeans came and to which they relayed back the results of their investigations. Phase-One science might be scattered around the globe, but only nations with 'a modern scientific culture', such as Britain, Holland and France, could 'fully appreciate, evaluate, and utilise' the knowledge thus acquired, though, in the course of assimilating new information from the wider world, Western science itself underwent modification.

In the second phase, that of 'colonial science', locally born or resident scientists (whom Basalla assumes to be Europeans) started to participate in scientific activities; local scientific institutions began to appear. While interest in natural history continued, almost all the scientific fields currently pursued in Europe were replicated overseas, but the local scientific community remained dependent upon European expertise and institutions and hence was reliant on 'an external scientific culture'. Basalla stressed that by calling colonial science 'dependent' he did not mean that it was necessarily inferior science (though critics have taken that to be his implicit meaning), and he claimed that the term could be applied not just to formal colonies like India, but also to science in uncolonised territories like China and Japan, or to the United States until several decades after its independence. The dependent status of colonial science ensured that many of its practitioners continued to receive their training in Europe and directed colonial scientists into areas of enquiry laid down by Europe. It remained difficult for colonial scientists to enter Europe's leading scientific societies and to gain access to those prestigious and influential 'invisible colleges' where the latest scientific ideas were debated and new agendas drawn up. The local scientific community had not yet reached the critical size necessary for 'reciprocal intellectual stimulation and self-sustaining growth'.

In time, as substantially larger numbers of scientists came to be trained and to work locally, extra-European societies in Phase Three strove to establish an 'independent scientific tradition' and a 'national science' of their own. Political independence might help to inspire greater scientific autonomy, but more important, Basalla averred, was the creation of national scientific institutions and honours. A political, educational and technological infrastructure emerged that allowed modern scientific research to thrive, conducted by local scientists operating within national boundaries and in accordance with national needs and priorities. Basalla suggested that the United States and Russia reached this stage between the two world wars, overtaking their former mentors in Western Europe. He placed Japan, Australia and Canada on a slightly lower stage of development, with other Asian, African and Latin American countries lagging well behind.

In the absence of a more satisfactory model, Basalla's typology has continued to be widely cited and discussed, though seldom with less than qualified approval. But Basalla was one of the first scholars to try to conceptualise 'colonial science', making it the transitional stage between the first implanting of 'modern science' overseas and its eventual maturation into 'national science' and an 'independent scientific tradition'. He did not, however, attempt to extend his analysis into medicine and technology (which in the case of nineteenth-century India might be seen as having far greater practical significance than science *per se*); nor, more puzzlingly, did he try to differentiate between varieties of colonialism. He ignored the enormous differences in background and experience of such countries as Brazil, Japan, India and Australia, and to anyone even superficially acquainted with their histories it would appear improbable that they all followed the same scientific trajectory through an identical time-scale. Leaving aside territories never formally colonised, there were clearly vast differences between the role of science in settler colonies, like those of North America and Australasia, where indigenous peoples and their cultures were largely swept aside, and 'colonies of exploitation' like India that were subject to colonial rule but where whites were few and indigenous cultures remained strong. By 'colonial science' Basalla clearly intended the science of itinerant and resident Europeans, though the term might equally apply (and has increasingly been applied) to that practised, in a colonial situation, by whites and indigenes alike. (It might also be noted, in passing and as an indication of the multi-stranded complexity of the scientific culture under discussion here, that even among European practitioners of science in a single colony there might be considerable diversity of origin and outlook. Although India's principal scientific and technological relations were inevitably with Britain as the colonial power, scientists from other countries – Danish botanists, German foresters, American malariologists – also contributed to the articulation of Western scientific ideas and practices in India.) Moreover, as Michael Adas points out, Basalla took a view, 'rarely challenged by his generation of scholars', that science was 'value neutral, objective, empirically demonstrated, somehow transcending time and thus universally valid'. Basalla's diffusionist vision was 'informed by a developmental teleology premised on the assumption that the spread of Western science to the rest of the globe [was] both beneficial and inevitable'.[22] This in turn led Basalla to assume that colonial regimes were willing agencies through which science could readily be diffused, whereas in many cases they might actually distort the development of science or, for ideological and material reasons, inhibit the spread of valued technologies. It might further be doubted whether, even after political independence, many erstwhile colonies had the resources to build their own 'national science' or to escape continuing dependence upon a small coterie of scientific superpowers.[23]

There are, of course, elements of the Basalla model that undoubtedly ring true. For instance, the emphasis given in Phase One to sciences such as botany and zoology is substantially borne out by Indian experience in the late eighteenth and early nineteenth centuries (though this disciplinary bias in fact persisted well into the twentieth century despite Indian attempts to redirect scientific enquiry to the 'purer' realms of mathematics, physics and chemistry). Basalla notes in his concluding remarks the need to examine science in different 'national, cultural, and social settings', but sadly he does not heed his own advice. Instead, by assuming that 'modern science' could have its origins only in the West and had therefore to be disseminated from there to hitherto 'non-scientific' countries, Basalla largely ignores the fact that countries like India and China had a long scientific and technological tradition of their own. Indigenous traditions did not simply evaporate with the first warming rays of an occidental sun: they were initially the subject of close, often appreciative, European scrutiny, and, though in India they were increasingly marginalised during the course of the nineteenth century, they continued to play a vital practical and ideological role. In Basalla's Eurocentric model, dynamism belongs to an (improbably) homogeneous West, leaving the rest of the world to participate only passively in the process of diffusion, unable to make any original contribution of its own or even to negotiate with an ascendant Western science.

---

[22] Michael Adas, 'Testing Basalla's Paradigm: Colonialism and the Diffusion of Science in British India', paper presented at the XIXth International Congress of the History of Science, Zaragoza, August 1993, p. 36.

[23] Nancy Stepan, *Beginning of Brazilian Science: Oswaldo Cruz, Medical Research and Policy, 1890–1920* (New York, 1976), pp. 14–40; Ian Inkster, 'Scientific Enterprise and the Colonial "Model": Observations on the Australian Experience in Historical Context', *Social Studies of Science*, 15 (1985), pp. 677–704; V. V. Krishna, 'The Colonial "Model" and the Emergence of National Science in India, 1876–1920', in Patrick Petitjean, Catherine Jami and Anne Marie Moulin (eds.), *Science and Empires: Historical Studies about Scientific Development and European Expansion* (Dordrecht, 1992), pp. 57–72.

INTRODUCTION: SCIENCE, COLONIALISM AND MODERNITY

Scientific knowledge and the ideology of science, it has been argued *contra* Basalla, can be 'actively redefined in the milieu of a recipient culture'. The receiving society, far from being supine, 'subverts, contaminates, and reorganises the ideology of science as introduced by Europe',[24] though one might add the caveat that the extent to which 'ideological subversion' could actually succeed in India before 1947 was constrained both by the political and financial control exercised by the colonial regime and by the influence and authority of the international scientific community.

In further refutation of an argument based on Western diffusionism and indigenous passivity, it is hard to see how, even at a superficial level, Western science could have functioned in many parts of the world without being able to draw upon 'local' knowledge and 'native' agency of various kinds, without local savants, scribes, interpreters and artists, fishermen and forest-folk, to guide and inform it. Increasingly, in conscious reaction against such ethnocentricity, many of the scientific 'discoveries' formerly claimed for the West have been traced back to earlier sources of indigenous knowledge. In the South Asia context, scholars have sought to establish the importance of Indian participation and information even in such seemingly exclusively European fields of colonial science as botany, geology and the trigonometrical surveys.[25]

But, if we are to reject a diffusionist model of 'colonial science', what can we put in its place? It could be argued that any attempt to reduce the complex experience of India to a simple typology is bound to fail, but there are at least two possible alternatives that deserve consideration. One is to argue that distinctions between centre and periphery, between 'metropolitan' and 'colonial' science, fundamentally misrepresent the way in which science evolved internationally from the seventeenth century onwards. Science, it might be reasoned, was not the property of a single society (located in Western Europe) but could be genuinely cosmopolitan, absorbing and assimilating information and ideas from a wide variety of sources and locations. 'Metropolitan science' did not even have a single, fixed locus of power: the 'metropolis' might move over time from one place to another; it might simply represent a certain way of doing or organising science, whether in Europe or overseas.[26] If we were to discard a

---

[24] Dhruv Raina and S. Irfan Habib, 'The Unfolding of an Engagement: *The Dawn* on Science, Technical Education and Industrialization, India, 1896–1912', *SH*, 9 (1993), pp. 87–8.

[25] E.g., Richard Grove, 'Indigenous Knowledge and the Significance of South-West India for the Portuguese and Dutch Constructions of Tropical Nature', in Richard H. Grove, Vinita Damodaran and Satpal Sangwan (eds.), *Nature and the Orient: The Environmental History of South and Southeast Asia* (Delhi, 1998), pp. 187–209.

[26] Roy MacLeod, 'On Visiting the "Moving Metropolis": Reflections on the Architecture of Imperial Science', *Historical Records of Australian Science*, 5 (1982), pp. 1–16.

Eurocentric approach, and jettison a constricted time-frame that privileges the eighteenth and nineteenth centuries, we might see that, through a long series of interactions, China, India, the Islamic world, even Meso-America, contributed as much, relatively speaking, to the development of science, technology and medicine before 1500 as Europe (and latterly North America) did over the following five hundred years. It could further be argued that, even within the colonial era, scientists in the colonies were equal participants, not inferior agents, in the development of science. Hence, a valid distinction cannot meaningfully be made between a 'low science' of fact-gathering in the colonies and a 'high science' of theory and synthesis in the metropole. Scientists in India, it can be argued, made major contributions to the natural sciences of the eighteenth and nineteenth centuries and won international recognition for their labours; 'colonial scientists' should not therefore 'be categorised as being separate from or inferior to the mainstream metropolitan scientific community'.[27]

However, although this line of interpretation helps to rescue many aspects of colonial science from apparent obscurity and metropolitan condescension, it does not necessarily take us much beyond the one-dimensionality of the diffusionist model. In particular it fails, much as Basalla does, to recognise the 'political character of science'. Although in some respects (in terms of the size of its scientific community, the number and quality of its societies and journals, its access to one of the West's premier scientific and industrial nations) India in the period under consideration enjoyed a relatively privileged position, it was still, none the less, a colonial, not an autonomous, domain. Whether under the East India Company or, following its demise in 1858, the Crown, there was a clear hierarchy of scientific authority that placed Britain and its scientists at the top and fixed India (and even British scientists working in India) in a position of dependence and subordination. Even in the closing decades of British rule, at a time when some historians have seen India as breaking free of imperial constraints in science, technology and medicine, as in many other spheres of activity, it is striking how entrenched British authority remained and how influential, despite the striving of 'national science', the models, agencies and agendas of metropolitan science remained. Science could not easily be divorced from the political ideologies and institutional structures that colonialism had put in place. If, as has aptly been said, Basalla's model 'trivialises the compulsions of colonialism',[28] it remains important to

---

[27] Satpal Sangwan, 'From Gentlemen Amateurs to Professionals: Reassessing the Natural Science Tradition in Colonial India, 1780–1840', in Grove et al. (eds.), *Nature*, p. 227.

[28] R. K. Kochhar, 'Science as a Tool in British India', *Economic and Political Weekly*, 17 August 1991, p. 1933.

INTRODUCTION: SCIENCE, COLONIALISM AND MODERNITY

keep those 'compulsions' in mind in seeking to devise other interpretative schema. Moreover, it is important to recognise that science, technology and medicine were more than a colonial force. They were, and surely remain, aspects of a global hegemony; it is prodigiously difficult for states, even those as large and powerful as India, even under Jawaharlal Nehru in the 1950s and 1960s, to attain their own scientific salvation. Internally, too, science became a vital factor in the articulation of class ideologies and structures, especially through the creation of novel systems of industrial organisation and production, through the creation of new professional and working-class identities and through the hegemonic authority of both established and aspiring elites. The term 'colonial science' (and its analogues in medicine and technology) may be flawed, but it is worth retaining and using it (more flexibly than Basalla did) to describe the various technologies of power operating within and through science in a colonial setting.

SCIENCE AS MODERNITY

Another way of approaching science and its significance in the context of colonial India is in terms of modernity. Science was intimately bound up with both colonial and Indian concepts of what constituted the modern world; it gave shape and authority to the understanding of modernity. By the early nineteenth century, following a period in which they showed themselves relatively receptive to Indian ideas and practices, the British saw science, technology and medicine as exemplary attributes of their 'civilising mission', clear evidence of their own superiority over, and imperial responsibility for, a land they identified as superstitious and backward. Science thus conceived served to heighten a growing sense of difference between Britain and India. In the wake of Britain's industrial revolution, technology (especially that of the steam age, heralded by steamships and railways) critically informed this perspective;[29] but so, too, did a growing sense of distinction between Western and indigenous medicine and a belief in the unique capacity of the West to master through engineering, medicine and natural science an environment that still held Indians in its thrall. Confidence in the transforming, modernising power of science climaxed with the viceroyalty of Lord Curzon (1899–1905), when the doctrines of high imperialism were echoed in the rhetoric and institutions of India's 'imperial science'. However, faith in Britain's capacity to modernise and civilise India was always fraught with multiple contradictions, among them a recurrent belief that

[29] Michael Adas, *Machines as the Measure of Men: Science, Technology and Ideologies of Western Dominance* (Ithaca, NY, 1989), chs. 3–4.

Indians were unready (or unfit) to receive the benefits of scientific modernity, a determination to deny India the competitive advantages that full access to modern science and technology might entail, and a romantically tinged anti-industrialism, in which India was destined to remain a land of princes, peasants and artisans, spared the ugliness and turmoil of modern industrial society.

Indians, unsurprisingly, also differed widely in their responses to Western science and its evangelising message of modernity. But some, especially among the Western educated, endorsed the call for India's transformation and identified wholeheartedly with the modernising project. As Gyan Prakash has put it, 'scientific reasoning became the organising metaphor in the discourse of the Western-educated elite. Impressed and stimulated by scientific and industrial progress in the West, the elite began to scrutinise indigenous religions and society in the light of scientific reason, not just rationality.' The 'cultural authority of science' and the 'authorisation of the elite as agents of modernity and progress' together attained 'an enduring dominance in India during the second half of the nineteenth century'.[30] As other scholars have shown, the reach of 'colonial modernity' extended well beyond institutional and economic reform to inform attitudes and practices relating to education and health, domesticity and gender roles, religious beliefs and social reform.[31]

But, for a colonial people, modernity could not be unproblematic. Modernity, and more restrictedly modernisation, has often been seen to represent the dissemination and acceptance of an essentially Western set of institutions and values, along much the same lines as Basalla's typology for the 'spread of Western science'. More than thirty years ago Lucien Pye defined modernisation as being 'based on advanced technology and the spirit of science, on a rational view of life, a secular approach to social relations, a feeling for justice in public affairs, and above all else, on the acceptance in the political realm of the belief that the prime unit of the polity should be the nation-state'.[32] Indians under colonial rule might endorse many items of this agenda but still feel that modernity remained for them an unattainable object of desire. Acceptance of modernity as partisanly presented by colonial officials, missionaries, educationalists and scientists would always confine

---

[30] Gyan Prakash, 'Science between the Lines', in Shahid Amin and Dipesh Chakrabarty (eds.), *Subaltern Studies IX* (Delhi, 1996), p. 60.

[31] Dipesh Chakrabarty, 'The Difference-Deferral of a Colonial Modernity: Public Debates on Domesticity in British India', in David Arnold and David Hardiman (eds.), *Subaltern Studies VIII* (Delhi, 1994), pp. 50–88.

[32] Lucien W. Pye, *Aspects of Political Development* (Boston, MA, 1965), p. 8; cf. S. N. Eisenstadt, 'Introduction: Historical Traditions, Modernization and Development', in S. N. Eisenstadt (ed.), *Patterns of Modernity II: Beyond the West* (London, 1987), pp. 1–11.

INTRODUCTION: SCIENCE, COLONIALISM AND MODERNITY

Indians to a state of tutelage and subordination, always leave them one step behind, second-best and imperfect copies of a Western ideal. How, Partha Chatterjee has asked, could Indians accept and assimilate the modernity of the colonising West while at the same time seeking to contest colonial authority and its assertions of Indian inferiority? He concludes that nationalism 'produced a discourse . . . which, even as it challenged the colonial claim to political domination, . . . also accepted the very intellectual premises of "modernity" on which colonial domination was based'.[33] Recent scholarship has sought to wrestle with this conundrum in various ways. One response is to argue, as Prakash does, that the authority of science had to be renegotiated and 'translated' to fit the needs and idioms of Indian society; it could not be accepted simply as it was articulated by the West. Another possibility is to suggest that modernity is not a single entity, patented by the West and retailed across the globe, but is capable of multiple forms and any number of cultural and political variants, which, while inevitably drawing on the science, technology and medicine of the West, also incorporate indigenous traditions and local systems of knowledge, thus enabling a country like India to forge a modern identity appropriate to its own cultural legacies and specific needs.[34]

As will be seen in this book, during the nineteenth and early twentieth centuries, Indian scientists and intellectuals tried to construct their own brand of Indian modernity, particularly through the selective incorporation (or re-invention) of Hindu ideas and traditions, though the mix of elements, the degree of 'hybridity' involved in this process, varied widely from one individual to another, even within the emergent scientific community. However, although this gave science in India a new sense of authority and belonging, it also generated its own dilemmas and sites of resistance. Even at the close of the colonial era it remained unclear how far scientific modernity could command a consensus within India itself and how far a science informed by Indian values could gain acceptance from an international scientific community dominated by the West. Whereas some nationalist politicians like Jawaharlal Nehru declared themselves to be worshippers at 'the shrine of science' and saw science as both the agency and emblem of Indian modernity,[35] others (led by the influential figure of M. K. Gandhi) spurned modernity, with its alien

---

[33] Partha Chatterjee, *Nationalist Thought and the Colonial World: A Derivative Discourse?* (London, 1986), p. 30.
[34] Prakash, 'Lines', pp. 61–2, 80; Partha Chatterjee, 'Our Modernity', in *The Present History of West Bengal: Essays in Political Criticism* (Delhi, 1997), pp. 193–210.
[35] Address to the National Academy of Sciences, Allahabad, 5 March 1938, in *Selected Works of Jawaharlal Nehru* VIII (New Delhi, 1976), p. 806; Jawaharlal Nehru, *An Autobiography* (London, 1936), pp. 433–7.

sciences and inappropriate technologies. To some extent these conflicting attitudes remain unreconciled in India today; as such they lie beyond the scope of this book.[36] But under Nehru, as independent India's first Prime Minister, a kind of pragmatic compromise was reached by which the nation-state assigned a no more than secondary role to the forces of tradition in science, technology and medicine, while identifying itself, and the needs of the people, with a more internationally recognisable brand of scientific and technological modernity.

[36] See Sunil Khilnani, *The Idea of India* (London, 1998).

CHAPTER 2

SCIENCE UNDER THE COMPANY

The English East India Company was as old as modern science itself. Founded in 1600, the Company shared its early years with the Scientific Revolution, and by 1662, when the Royal Society of London was founded, was already a flourishing concern with trading bases at Surat, Madras and Masulipatam. The sciences prominent in early colonial India – botany, geology, to a lesser extent zoology – were still at a formative stage when the Company embarked on its career of territorial expansionism in the mid-eighteenth century. The first volume of Buffon's *Histoire naturelle* appeared in 1749, as Anglo-French rivalry in the Carnatic was reaching its peak; Linnaeus's *Species plantarum*, which established the binomial system of nomenclature, was published in 1753, four years before the battle of Plassey opened the floodgates to British ascendancy in Bengal. By the time the Geological Society of London, model for a new generation of metropolitan scientific societies, was founded in 1807, British power had been extended over vast tracts of northern and peninsular India and was poised for the final defeat of the Marathas. The publication in 1830 of the first volume of Lyell's *Principles of Geology*, one of the foundational texts of modern geology, came three years before the Company lost its vestigial trading rights; and Darwin's *Origin of Species* was published in November 1859, twelve months after the East India Company had finally been declared extinct.

Thus Company rule in India was contemporaneous with one of the most momentous phases of modern science, from the rise of Enlightenment natural history to the eve of Darwinian biology. The Company was accordingly well placed to contribute to, and to profit from, the development of science, but how central was science to the Company? Was it more decorative or symbolic than functional in its significance? What kind of science flourished under the Company? Who practised science and why? How was India represented in and through the science of the period? These are questions this chapter will attempt to explore.

THE COMPANY OF SCIENCE

The East India Company's Court of Directors in London exercised a commanding position in relation to science in India. One of the leading patrons of science in Britain itself, on the subcontinent the Company and its servants enjoyed a near monopoly over Western scientific activity. Anxious to preserve its commercial privileges and prevent outsiders from undermining its authority, the Company closely regulated European access to India. Its approval was essential for any kind of scientific expedition to be undertaken and the Company was disinclined to allow scientific visitors, however eminent they might be. Apart from the French naturalist and traveller Victor Jacquemont, who died in India in 1831, the greatest exception to the scientific monopoly of Company servants was the expedition to India in 1848–50 of **Joseph Dalton Hooker**,[1] the foremost botanist of nineteenth-century Britain. The German naturalist Alexander von Humboldt sought, but was never granted, permission to visit India.[2] Many leading British scientists of the period – Joseph Banks, Charles Lyell, and Charles Darwin among them – showed great interest in the natural history of India without ever visiting the country in person. Europe's scientists and collectors relied instead on informal networks of contacts with army officers, doctors and officials – or on the magnanimity of the Company itself – to provide them with specimens, drawings and scientific information.[3]

Through its control of the nomination of candidates for bureaucratic and military appointments in India, the Court of Directors commanded a vast fund of patronage and regulated access to one of the richest fields of scientific and technical employment available to Britons overseas. Even once taken into the Company's employment, those of its servants who wished to pursue scientific interests remained heavily dependent on the approval and funding of the Court of Directors or its most senior representative in India, the Governor-General. In the 1790s, in an early instance of its support for science, the Company paid £300 for the surgeon-botanist **William Roxburgh** to begin publication of his lavishly illustrated *Plants of the Coromandel Coast*. By the time of its completion in 1819, the three-volume work had cost the Company £2,000, but by then the market for such costly works of natural history had already collapsed. This 'commercial disaster' did much to discourage the Company from further acts of largess, to the dismay of later naturalists, who believed themselves entitled to similarly generous treatment.[4] In the 1840s, the

---

[1] Information on individuals whose names are in bold will be found in the Biographical Notes at the end of this book, pp. 214–16.

[2] Jean Théodoridès, 'Humboldt and England', *BJHS*, 3 (1966), pp. 43–4.

[3] Apart from Hooker, Darwin's India informants included three Company surgeons (J. G. Malcolmson, J. Forbes Royle and Hugh Falconer), a Madras civil servant, Walter Elliot, and the curator of the Asiatic Society's museum, Edward Blyth: Frederick Burkhardt and Sydney Smith (eds.), *The Correspondence of Charles Darwin*, vols. II–VII (Cambridge, 1986–91).

[4] Ray Desmond, *The European Discovery of the Indian Flora* (Oxford, 1992), pp. 47–50; Simon Schaffer, 'Visions of Empire: Afterword', in David Philip Miller and Peter Hanns Reill (eds.), *Visions of Empire: Voyages, Botany and Representations of Nature* (Cambridge, 1996), p. 343.

Company subsidised publication of **Hugh Falconer** and **Proby T. Cautley**'s catalogue of the Siwalik fossils from north India, promising to take forty copies at a guinea each. But, despite lobbying from the Royal Society and other scientific bodies in London, the Directors declined to pay for the full cost of classifying, illustrating and exhibiting the 40 tons of fossils Falconer and Cautley had shipped back from India. Forty guineas was a modest sum compared with the £1,000 donated by the British government.[5] The Court of Directors showed little interest, too, in giving Hooker the financial support he expected to publish his *Flora Indica*, the first and only volume of which appeared in 1855; the full publication of the single most important work of nineteenth-century Indian botany had to wait until after the abolition of the Company to find administrative favour and a suitable subsidy. It was not surprising, therefore, that, despite the effusive dedications and public tributes to the Company's 'princely' patronage, in private many men of science railed against the 'scoundrely' and 'knavish' conduct of those 'cheese-monger Emperors', the Court of Directors.[6]

Some historians have been at pains to show that the Company's scientific interests were not determined by purely material considerations but represented a more enlightened and disinterested approach to science. While accepting that the 'profit motive' necessarily 'guided the policies of the East India Company', Ray Desmond has contested the kind of argument that reduces all the decisions and actions of the Court of Directors to 'solely ... commercial considerations'. He points out that the Directors 'founded and maintained ... an unremunerative library and museum in Leadenhall Street in the City, were generous patrons of scholarly publications, and seldom opposed their officials pursuing Oriental studies'.[7] But, since the Company was under frequent attack for mismanaging Indian affairs, and given that science was a prestigious pursuit among Britain's ruling classes, it must, at the least, have been politically advantageous to be hailed as one of the 'most munificent patrons of science in the world'. It may have been the case that, conveniently, 'self-interest and scholarship often coincided',[8] but whereas the former was frequently in evidence, the latter, at best, received only erratic support.

From time to time the Court of Directors, or its principal representatives in

---

[5] Jack Morrell and Arnold Thackray, *Gentlemen of Science: Early Years of the British Association for the Advancement of Science* (Oxford, 1981), p. 352.

[6] Leonard Huxley, *Life and Letters of Sir Joseph Dalton Hooker*, 1 (London, 1918), p. 358; Marika Vicziany, 'Imperialism, Botany and Statistics in Early Nineteenth-Century India: The Surveys of Francis Buchanan (1762–1829)', *MAS*, 20 (1986), p. 626, note 5.   [7] Desmond, *Discovery*, p. v.

[8] T. J. Newbold, 'Summary of the Geology of Southern India', *JRAS*, 12 (1850), p. 95; Mildred Archer, 'India and Natural History: The Role of the East India Company, 1785–1858', *History Today* (November 1959), p. 736.

India, declared their high-minded attachment to science, recognising its promotion as the responsibility of a civilised government. Thus, in 1804, at a time of growing British ascendancy, the Governor-General, Lord Wellesley, declared that 'to facilitate and promote all enquiries which may be calculated to enlarge the boundaries of general science' was a duty 'imposed on the British Government in India by its present exalted situation'. Yet, as Marika Vicziany has shown in discussing the career of the naturalist **Francis Buchanan**, Wellesley was well aware of the value that scientific surveys might have in advancing his reputation and deflecting criticism of his expansionist policies.[9] Science could also serve more narrowly utilitarian ends. In 1787 a Polish botanist, Anton Hove, was sent by the Board of Trade in London (for once without Company approval) to report on cotton-growing and textile production in Gujarat, an area still under Maratha control. Disguised as an indigent physician, Hove reported extensively on agriculture, medicine and other matters of scientific and technological interest and shipped back several crates of plants for the botanic gardens at Kew, but the commercial and political motives behind his mission were ill concealed.[10]

As British power in South Asia grew, its human and material resources attracted the close attention of a revenue-hungry administration, and several scientific surveys were commissioned to provide the Company with more information about its newly acquired territories. Among the most important of these were the survey of Mysore under Colonel **Colin Mackenzie** between 1799 and 1810, the journey of Francis Buchanan from Madras through Mysore to Kanara in 1800–1, and Buchanan's survey of Bengal and Bihar, 1807–14. As with the scientific activities of the Company in general, these surveys have been the subject of growing scholarly debate. Were they driven by purely material considerations (arising from the Company's need to 'know' India the better to rule and exploit it), or did they express a wider vision of scientific needs and opportunities? Were they simply instruments in the imperial ordering of India, or did they represent the more explicitly scientific objectives of Company servants themselves? The issue is not easily resolved, for the surveys served a variety of purposes and the motives of individuals like Mackenzie and Buchanan were not necessarily those of their paymasters. In describing his own objectives for the Mysore survey, Mackenzie declared in 1800 that his intention was to 'obtain . . . a clearer and better

---

[9] Deepak Kumar, *Science and the Raj, 1857–1905* (Delhi, 1995), p. 70; Vicziany, 'Imperialism', pp. 627–38.
[10] A. P. Hove, *Tours for Scientific and Economical Research Made in Guzerat, Kattiawar, and the Conkuns, in 1787–88* (Bombay, 1855).

defined knowledge of the extent, properties, strength and resources' of the territory recently seized from the defeated state of Mysore. This would help to provide commercial intelligence for the Company, but it would also illuminate 'many objects of natural history' that would contribute to 'the improvement of scientific knowledge'.[11]

The directions given to Buchanan by the Court of Directors in 1807 for his survey of eastern India also combined matters that were directly relevant to the Company's financial interests with those that would appear to be of purely scientific or curiosity value. Buchanan was asked to report on the topography and natural resources of each district, its 'extent, soil, plains, mountains, rivers, harbours, towns and subdivisions', together with the 'air and weather', and whatever he might discover 'worthy of remark concerning the history and antiquities of the country'. With respect to 'the natural productions of the country', he was to enquire into the nature of animal, vegetable and mineral products, especially those used as food and medicine or in trade and manufacture; he was to enquire about fisheries, forests, quarries and mines. Agriculture, too, was to receive attention – from crops, implements, and livestock to the general state of farms and landed property. He was further to report on the inhabitants of the region, their customs, commerce and manufacturing. Finally, he was to 'take every opportunity of forwarding to the Company's Botanical Garden [in Calcutta] ... whatever useful or rare or curious plants and seeds' he discovered and such 'observations' as might be necessary for their cultivation. The phrase 'useful or rare or curious' aptly conveys the intermingling of the functional and the ornamental in early Company science, though it is clear that the matters Buchanan was most persistently called upon to investigate were those most likely to prove of material worth.[12] However, Buchanan's personal preference, like that of many other Company servants, was for natural history, above all his 'beloved botany'. As Vicziany points out, in his surveys Buchanan was greatly influenced by the model of Sir John Sinclair's *Statistical Account of Scotland* and its underlying doctrine of 'improvement', a term that repeatedly surfaces in Buchanan's account of war-ravaged Madras and Mysore.[13] With their concern for wealth, resources, and means of subsistence, for population, climate, irrigation, and customs, these early surveys might best be understood as early essays in what Michel Foucault called 'governmentality', though this was punctuated, in Buchanan's case, by

---

[11] Kumar, *Science*, p. 33; Peter Robb, 'Completing "Our Stock of Geography", or an Object "Still More Sublime": Colin Mackenzie's Survey of Mysore, 1799–1810', *JRAS*, 8 (1998), pp. 181–206.
[12] Francis Buchanan, *An Account of the Districts of Bihar and Patna in 1811–1812* (Patna, 1934), pp. i–iv.
[13] Vicziany, 'Imperialism', pp. 643–4, 648–9.

outbursts of a Romantic sensibility to landscape, most evidently in his descriptions of waterfalls and forests.[14]

Although the surveys of Mackenzie and Buchanan have attracted considerable historical attention and demonstrate the extent to which the Company and its servants at times sought systematically and in a self-consciously scientific way to map and compile inventories of newly acquired territories, they were to some extent exceptional for the period. In the main the Company did not initiate scientific projects, but merely tolerated its employees' private pursuits. Army officers, civil servants, engineers and physicians might satisfy their cultured curiosity or while away their leisure hours by keeping meteorological records, hunting for fossils, or compiling notes on local flora and fauna, but that did not necessarily interest or concern their superiors in Calcutta or London. Indeed, the nature of Company rule and the burden of official duties might do more to obstruct than to facilitate the pursuit of 'recreational' science. Few Europeans in India had the time and opportunity to dedicate themselves wholeheartedly to science. 'A mere man of letters, retired from the world and allotting his whole time to philosophical or literary pursuits' was said in 1788 to be 'a character unknown among Europeans resident in India'. Everyone was 'constantly occupied either in the affairs of Government, in the administering of justice, in some department of revenue or commerce, or in one of the liberal professions'.[15] Unlike in Britain, there were few European clergymen, landlords and manufacturers to produce papers on natural history or patronise fledgling scientific societies. A want of time for 'general researches' dogged even those in the Company's medical service, like the botanist Roxburgh, and constituted the principal 'obstacle to the progress of knowledge' as they understood it.[16] Several among those who had begun to establish scientific reputations died young or returned to Britain and so were lost to India. Rarely did European men of science retire in the country. **B. H. Hodgson**, who wrote prolifically on Himalayan zoology and ethnography between the 1820s and his eventual return to Britain in 1858, was a rare example of a naturalist who elected to stay on, following his resignation from government service in 1843.[17]

---

[14] Michel Foucault, 'Governmentality', in Graham Burchell, Colin Gordon and Peter Miller (eds.), *The Foucault Effect* (Chicago, 1991), p. 93; Francis Buchanan, *Journey from Madras through the Countries of Mysore, Canara, and Malabar* (London, 1807), I, pp. 126, 170; II, pp. 24, 166–7. For tensions between these ideas in Buchanan's homeland, see Peter Womack, *Improvement and Romance: Constructing the Myth of the Highlands* (Basingstoke, 1989).
[15] Rajendralal Mitra, 'History of the Society', *Centenary Review of the Asiatic Society of Bengal from 1784 to 1883* (Calcutta, 1885), p. 18.   [16] Kumar, *Science*, p. 66.
[17] William Wilson Hunter, *Life of Brian Houghton Hodgson* (London, 1896).

Far from embodying the rule of science in the service of empire, the Company's involvement in science often appeared more like a fitful flirtation. From time to time, when its economic and political interests were aroused, or the case for some scientific endeavour was convincingly made, science could command the Company's active attention, but for much of the period it was of secondary importance compared with the more pressing concerns of revenue, diplomacy, law and order. The British enjoyed the company of science; it would be excessive to suggest that they ruled by it. A large part of the science conducted in India before 1858 accordingly belonged not to the state but to the tradition of 'gentlemanly' science that flourished in Britain until the mid-nineteenth century. It mostly lacked a specialist state agency and a clearly defined agenda. Those who spoke for India's science, whether in Britain or in South Asia, frequently found the Court of Directors and the Government of India perplexingly uninterested, even on matters that seemed of material value to the Company. The lack of official support for an Indian branch of the British Association for the Advancement of Science in the 1830s was indicative of this apparent unconcern.[18] In the circumstances, as H. J. C. Larwood observed, 'the enthusiasm with which scientific interests were pursued and the bulk of work produced was remarkable', though recent scholarship is far less disposed to accept his conclusion that much of the work done by scientists in the Company period was 'of little permanent value'.[19]

Until the 1840s and 1850s, army engineers (such as Mackenzie) and surgeons (such as Buchanan) were called upon to furnish the administration with all-purpose expertise without the expense and delay of recruiting specialists from abroad. Only towards the end of the Company era, as scientific disciplines grew more specialised and the requirements of scientific administration more taxing, were outsiders recruited for Indian service. In an innovative move, in 1851 Thomas Oldham, previously Director of the Geological Survey of Ireland, arrived to take charge of what five years later became the Geological Survey of India. But even then the resources placed at his disposal – an office, a clerk and a *chaprassi* – were far from adequate for the task in hand and suggested the Company's still limited practical commitment. With the partial exception of the Great Trigonometrical Survey of India, which dated from 1818 but again drew on the army for its personnel, dedicated scientific and

[18] Morrell and Thackray, *Gentlemen*, pp. 502–3.
[19] H. J. C. Larwood, 'Western Science in India before 1850', *JRAS*, Part 2 (1962), p. 76; cf. Satpal Sangwan, 'From Gentlemen Amateurs to Professionals: Reassessing the Natural Science Tradition in Colonial India, 1780–1840', in Richard H. Grove, Vinita Damodaran and Satpal Sangwan (eds.), *Nature and the Orient: The Environmental History of South and Southeast Asia* (Delhi, 1998), pp. 210–36.

technical agencies had barely begun to emerge from the matrix of the military and medical services before the 1850s. It is necessary, therefore, to look beyond the formal parameters of state science to establish the general character and significance of science in the Company period.

THE ORGANISATION OF COLONIAL SCIENCE

Much of the scientific endeavour of the Company period took place outside, or on the margins of, state institutions. Many early accounts on Indian geology, for instance, were written by military officers and army surgeons not as part of their official duties but under the stimulus of personal interest and in the course of cross-country marches. Thus, Captain Franklin of the Bengal Army compiled his pioneering observations on the geology of central India during a march through Bundelkhand in 1826–7; Assistant Surgeon Hardie slipped away from his regiment while marching from Baroda to Udaipur to examine wayside rocks, later apologising to his readers for not having been able to carry out more systematic investigations. As a by-product of troop movements, such accounts were bound to be superficial, but they sketched in the first outlines of Indian stratigraphy.[20]

Some of the most significant discoveries of the period were made without official sponsorship, or even much prior knowledge of a particular science. A critical example was the unearthing of the Siwalik fossils in November 1837, a momentous event for Indian palaeontology and for wider discussions of climatic change and extinct mammalian species. Hugh Falconer, a young Company surgeon and Superintendent of the Saharanpur Botanic Garden, followed up earlier finds made in October 1834 by Lieutenant Proby T. Cautley of the Bengal Engineers and his assistants W. E. Baker and H. M. Durand while excavating the Jumna Canal. As in Britain, major engineering works of this kind created unique opportunities for geological discoveries, but Falconer was also the beneficiary of indigenous knowledge: the local raja, who possessed a huge mastodon molar (known as 'the tooth of Deo'), advised him where to dig for further specimens and in six hours 300 fossil bones were uncovered.[21]

The discovery and identification of the Siwalik fossils established the scientific reputations of Falconer and Cautley in Europe, and they became joint recipients of the Geological Society's prestigious Wollaston Medal in

---

[20] James Franklin, 'On the Geology of a Portion of Bundelkhand, Boghelkhand, and the Districts of Sagar and Jebelpur', *AR*, 18 (1833), pp. 23–46; James Hardie, 'Remarks on the Geology of the Country', ibid., pp. 82–99.

[21] E. W. C. Sandes, *The Military Engineer in India*, II (Chatham, 1935), pp. 275–7.

1837. Cautley returned to his canal, Baker and Durand's official duties left them little time for amateur geology, but Falconer, still under 30, became an influential figure in British geology. Although obliged to return to India, he was elected a Fellow of the Royal Society in 1845 and served as a member of its Council and as Vice-President of the Geological Society until his death in 1865. Falconer's career (like that of his contemporary, the surveyor **George Everest**) demonstrated how India could provide a springboard for a metropolitan reputation and its accompanying honours.[22] Andrew Grout has argued that the award of the Wollaston Medal had 'great symbolic value', not just for Falconer and Cautley, but for all geologists working in India. 'It showed that the field was still open to enthusiastic amateurs, and it was taken as evidence that the geology of India, and those labouring in its elucidation, were considered important by metropolitan savants.'[23] And yet it was suggestive of the relative backwardness attributed to colonial science that, in presenting the medal in 1837, Lyell pointed out that when Falconer and Cautley made their discovery they were not 'versed in fossil osteology' and, 'being stationed on the remote confines of our Indian possessions', were far from any 'living authorities' or standard work of palaeontology. Lyell commended 'the manner in which they overcame these disadvantages, and saw the enthusiasm with which they continued for years to prosecute their researches, when thus isolated from the scientific world' as being 'truly admirable', but the imputation was, none the less, that science in the colonies was inevitably a more amateurish pursuit than in the metropole.[24] Falconer was himself loath to leave London to resume his colonial 'isolation', and Darwin and Hooker, too, regarded his return to India as a significant loss to metropolitan science. Once back in Calcutta, Falconer showed his displeasure by being petulant and snobbish towards the Asiatic Society, the body that had first brought the Siwalik fossils to international attention.[25]

Whether justified or not, an acute sense of isolation haunted the geological science of the Company period. When combined with British empiricism, it bred a science that was as diffident as it was dependent. The surgeons, engineers and army officers who first wrote about geology often lamented their

---

[22] Horace B. Woodward, *The History of the Geological Society of London* (London, 1907), pp. 128–9, 199; Matthew H. Edney, *Mapping an Empire: The Geographical Construction of British India, 1765–1843* (Chicago, 1997), p. 147.
[23] Andrew Grout, 'Geology and India, 1770–1851: A Study in the Methods and Motivations of a Colonial Science', Ph.D., University of London, 1995, p. 83.
[24] P. N. Bose, 'Natural Science', in *Centenary Review*, p. 61.
[25] Hooker to Darwin, 25 March 1846, in Burkhardt and Smith (eds.), *Correspondence*, III, p. 305; Hooker to Darwin, 13 October 1848, *Correspondence*, IV, p. 172.

own lack of expertise. They saw themselves as amateurs, essentially confined to 'collecting and recording with precision'; this was (to their minds) a worthy enough task but its main function was to provide the raw data that more expert minds in Europe might evaluate and incorporate into wider systems of scientific knowledge. 'I am not learned enough in the science to become an advocate for any party', observed Franklin, alluding to the Huttonian–Wernerian controversy that had dogged British geology earlier in the century; 'a few lessons when in England, and the great volume of nature have been my chief guide'. All he aspired to was 'to record facts, to lay down strata correctly though not minutely on a map, [and] to extract that which is useful or profitable in the science'.[26] There was a feeling, too, not confined to geology, that even the most talented scientific minds in India failed to attain their full potential or receive the recognition they deserved. The *Calcutta Review* claimed that James Prinsep, who wrote more than sixty articles for the Asiatic Society between 1825 and 1839 on subjects as diverse as meteorology, geology and numismatics, could have become 'the Humboldt of the east' had he only 'been spared to us', but he died in 1840, aged only 41. Despite the zeal and energy displayed by Company scientists, their talent, it seemed, was spread thinly and over too many fields, or official duties prohibited a more single-minded pursuit of science. Even those who gained a reputation outside India – such as Falconer, Everest, and the surgeon-botanist **J. Forbes Royle** – felt like novices when they first ventured into scientific circles in Britain.[27] Nor was the isolation that Lyell referred to much diminished by the end of the Company period. 'We labour under immense disadvantages in this country in the want of books of reference and collections for comparison', remarked Thomas Oldham in 1856. 'We are working under the same difficulty that would affect a tradesman without his tools or a physician without medicines.' In these circumstances it was 'utterly impossible to attain the same progress which should be looked for in Europe'.[28]

Not all science under the Company was as itinerant as Buchanan's surveys or as fortuitous as finding the Siwalik fossils. Up-country surveys, missions of scientific reconnaissance and opportunistic observations during troop marches and canal works were supplemented by the science of city-based learned societies, journals and museums, and these, as much as official policy and patronage, helped shape the character of colonial science. The earliest and most celebrated of India's learned societies, the Asiatic Society of Bengal, was

---

[26] Franklin, 'Geology', p. 37.   [27] 'Works on the Himalayas', *CR*, 4 (1845), pp. 163–5.
[28] Satpal Sangwan, 'Reordering the Earth: The Emergence of Geology as a Scientific Discipline in Colonial India', *IESHR*, 31 (1994), p. 305.

established in Calcutta in January 1784 on the initiative of the judge and Orientalist Sir William Jones. Though inspired by the Royal Society in London (one early indication of the role that the Royal Society was to play as model and mentor for science in India throughout the colonial period), the Asiatic Society was not primarily a scientific body.[29] However, in the absence of the kinds of specialised scientific societies that flourished in London after the founding of the Geological Society in 1807, it functioned as the main platform for Western science in India until replaced by the Indian Science Congress in 1914. Articles on scientific topics appeared intermittently in the Society's *Asiatic Researches* between 1788 and 1839, but, like the Company itself, it showed greater favour towards literary and linguistic studies. In 1808, in an attempt to encourage more scientific contributions, a committee was established 'to promote the knowledge of natural history, philosophy, medicine, improvements of the arts and sciences, and whatever is comprehended in the general term physics'.[30] The 'Physical Class', though revived in 1828, met only erratically and attendance was poor. By the 1830s the Asiatic Society of Bengal was competing with other scientifically oriented Calcutta societies, including the Agricultural Society set up in 1820 and the Medical and Physical Society in 1823. Beyond Bengal, comparable societies also came into existence, notably the literary, scientific and medical societies of Bombay and Madras.[31] The Royal Asiatic Society in London, founded in 1823, provided a further forum for the presentation and discussion of Indian material and, by virtue of its location, was more closely integrated into the world of metropolitan scientific societies.

Museums, too, were important not just as sites for the pursuit of scientific knowledge but as a way of establishing the reputation and self-esteem of colonial science. Although vast quantities of botanical, zoological and geological specimens were sent to London to form part of the Company's India Museum or for distribution to Continental collections, many were retained in India. As the region's premier scientific body, the Asiatic Society was anxious that its collection (including thousands of rock and mineral specimens) should be properly housed and displayed and that Calcutta should have a museum that would vie with institutions in Europe and yet be a scientific window onto India. First proposed by the naturalist **Nathaniel Wallich** in 1814, the museum was intended 'for the reception of all articles that may tend to illustrate Oriental

---

[29] O. P. Kejariwal, *The Asiatic Society of Bengal and the Discovery of India's Past, 1784–1838* (Delhi, 1988).
[30] 'Introduction', *AR*, 18 (1833), p. i.
[31] A. Neelameghan, *Development of Medical Societies and Medical Periodicals in India, 1780–1920* (Calcutta, 1963).

manners and history, or to elucidate the peculiarities of art and nature in the East'.[32] The Bombay Literary Society, founded in 1804, followed suit with its own museum of natural history and antiquities in 1815. Museum collections had their practical uses. In 1835 the Government of India, enticed by prospects of exploiting coal and other minerals, decided to found its own museum of economic geology, initially entrusting the task to the Asiatic Society before, in 1856, transferring the collection to the Geological Survey.

Museums were also one of the institutional sites on which a more professionalised science was beginning to be established. In 1841 the zoologist Edward Blyth was sent out from England as the Society's full-time museum curator, a post he held until 1862. His appointment, paid for by the Company, has been seen as initiating a new phase in the history of colonial science in India, but his predicament was also indicative of the constraints on scientific endeavour in this period. As a Company employee, Blyth was tied to his official duties and needed permission even to leave Calcutta. In fourteen years, he lamented to Darwin in 1855, he had 'hardly been out of' the city and lacked the 'great advantages of travel and personal observation elsewhere'. Poor pay and the want of books, space and competent assistants left Blyth desperate to return to England.[33]

In an age in which the printing press assumed an indispensable role in the dissemination of scientific ideas,[34] journals, too, had a prominent place in nurturing Western science in India. Not only did they emulate the practice of learned societies in Europe; they also helped to maintain contact between educated Europeans scattered across India and adjacent territories, and served as a means by which scientific observation and speculation could be relayed to and from India. In 1829 Captain J. D. Herbert, Deputy Surveyor-General, launched the monthly journal *Gleanings in Science*. Intended as both a digest of scientific articles from Europe and a vehicle for local communications, *Gleanings* was soon swamped by the latter, which ranged with little discrimination from the introduction of steamboats on the Hooghly to ways of keeping wine cool in India. One reviewer complained that, despite the wealth of opportunity India presented for scientific study, the Asiatic Society had neglected physical science and natural history; Jones's interests had been predominantly literary and the government of a foreign country was bound to attach more importance to language than to science.[35] No less discouraging

---

[32] B. V. Subbarayappa, 'Western Science in India up to the End of the Nineteenth Century AD', in D. M. Bose, S. N. Sen, and B. V. Subbarayappa (eds.), *A Concise History of Science in India* (New Delhi, 1971), pp. 528–9.   [33] Burkhardt and Smith (eds.), *Correspondence*, v, pp. 392, 401.
[34] Cf., John Ziman, *The Force of Knowledge: The Scientific Dimension of Society* (London, 1976), pp. 97–9.
[35] 'Analyses of Books', *Gleanings*, 1 (1829), p. 211.

than the indifference of the state was the apathy of the public. At a time when the Geological Society in London could boast of having more than 700 members, membership of the Asiatic Society seldom exceeded 250: in 1845 it sank to a miserable 119. The weekly sessions with which the Society began had dwindled by 1799 to monthly meetings. India's other learned societies could expect even fewer supporters and subscribers. In 1825 the Medical and Physical Society of Calcutta had just over 200 members. In 1829 the editor of *Gleanings* estimated that the journal-reading public in Bengal numbered no more than 2,000. Three years later, *Gleanings* was taken over by the Asiatic Society and incorporated into the *Journal of the Asiatic Society*, which over the next fifty years published several hundred papers on scientific subjects.[36]

Undeterred by the fate of *Gleanings*, several other scientific and medical journals sought to establish themselves between the 1820s and 1850s. They included the *Calcutta Journal of Natural History*, launched in 1840 by Surgeon John M'Clelland, who wrote many of the early articles himself. His opening editorial was not auspicious: 'The usual apology for being without a periodical in the metropolis of British India exclusively devoted to objects of science is that it would not pay.' He was determined, none the less, to put Calcutta's 'taste to the test'.[37] He also hoped the *Calcutta Journal* would become the focus for an Indian branch of the British Association for the Advancement of Science, believing a periodical especially necessary for this purpose in India, where 'the cultivators of science are so few, and the nature of their various duties such as to prevent their assembling at any one place'.[38] But the idea failed to win support and the journal folded in 1847.

Many of the journals of the period set out to educate and inform their readership in ways that suggest the importance of science less as an instrument of state than as a means of promoting the cultural cohesion and social identity of the European elite. It has been argued that in an increasingly class-riven Britain scientific societies helped give a sense of solidarity and collective purpose to the new, urban middle class: science symbolised order in a world threatened by rapid social change and working-class militancy.[39] Matthew Edney has similarly argued that the pursuit of science in India helped to differentiate the European elite from both Indians and British soldiers, merchants and planters.[40] Science certainly served to promote and preserve a sense of cultural commonality among educated Europeans in India while helping them to maintain contact with cultured society in Europe. It reassured the colonial elite that they

---

[36] Subbarayappa, 'Science', p. 500. For further discussion of the nature and significance of scientific publications in this period, see Grout, 'Geology and India', pp. 87–104.
[37] *CJNH*, 1 (1840), p. 1.   [38] Ibid., p. 8.   [39] Morrell and Thackray, *Gentlemen*, pp. 22–31.
[40] Edney, *Mapping*, p. 33.

belonged to a world of metropolitan taste – or could gain ready access to it. The editor of a volume of *Asiatic Researches* devoted to geology maintained that every European in India was capable of making some contribution to science. 'Scattered as are our countrymen in the East, over so large a portion of the surface of the earth as yet unexplored by science, the most common observer can hardly fail to notice phenomena that may be important for the purpose of physical research.' The 'unscientific enquirer' might do little more than collect and observe, but he (the assumption was invariably masculine) could pass his data on to others better able to use them, or, through self-instruction, he could readily acquire a 'more accurate knowledge' for himself.[41] The *Calcutta Review*, launched in 1844, also took its didactic responsibilities seriously. Its declared intention was to bring together 'useful information' and 'sound opinions' about Indian affairs, and to 'conduce ... directly or indirectly, to the amelioration of the condition of the people'. 'The bane of this country is ignorance', the editor declared, not the ignorance that existed in the 'dark recesses of native life', but that which existed 'in high places – among the ruling body – among the men to whom inscrutable Providence has submitted the destinies of India'.[42]

Implicit here, as in much of the scientific literature of the period, was criticism of the Company for not being better informed about, or more receptive to, science. But the *Review*'s passing reference to 'the dark recesses of native life' highlights another aspect of European science under the Company – its externality. Despite declarations, like that made by the Governor-General, Lord Bentinck, in 1835, that the 'great object' of the British government in India 'ought to be the promotion of European literature and science among the natives of India',[43] when it came to scientific societies, museums and journals, it was evident that Western science was primarily about India and Indians, and only secondarily for and by them. The founding members of the Asiatic Society were all Europeans, and Jones in his inaugural address left open for later consideration whether Indians should be allowed to join. Though papers were from time to time submitted and read on their behalf by European members, no Indians were admitted to membership of the Society until 1829. Thereafter they were actively involved, including Ramkamal Sen, who served from 1838 as its 'native secretary' and donated many items to the Society's museum. In its early years, too, the Bombay Literary Society had as much a social function as an intellectual agenda: 'Its members formed a select

[41] *AR*, 18 (1833), p. ii.   [42] *CR*, 1 (1844), p. ii.
[43] John Rosselli, *Lord William Bentinck: The Making of a Liberal Imperialist, 1774–1839* (London, 1974), p. 220.

circle of high-placed Europeans who met together for mutual friendship as well as for "the feast of reason and flow of soul".' When, after 1840, as a 'natural consequence of the growth of education in the country', Indians began to be admitted, 'it involved a considerable departure from the original design'.[44]

It would, nevertheless, be a mistake to see Indians as totally isolated from the scientific enterprise of the period. A striking example of their direct engagement was the career of Bal Shastri Jambedkar, a Brahmin from Ratnagiri district, who became secretary to the Bombay Educational Society in 1827 at the age of 17 and translated Lord Brougham's *Treatise on the Objects, Advantages and Pleasures of Science* into Marathi. He became 'native secretary' to the Oriental Translation Committee of the Bombay branch of the Royal Asiatic Society, working alongside British Orientalists. When Elphinstone College was established in 1834 he became a professor of mathematics and, as well as teaching astronomy and mathematics, translated scientific works into Marathi. He also encouraged Indian interest in chemistry, natural history and other aspects of the 'useful' knowledge of the West through his Anglo-Marathi journal, the *Darpan*. The journal, and its successor *Dig Darpan*, reflected the attitudes and concerns of the first generation of Western-educated in Bombay, 'convinced of the need for an expansion of India's intellectual horizons and for a revaluation of her traditional social and religious beliefs'.[45] Perhaps because it had a more practical programme than the Asiatic Society, Calcutta's Agricultural and Horticultural Society had from its earliest years Indian officers and members, including such leading *bhadralok* figures as Dwarkanath Tagore and Radhakanta Deb. In keeping with its wider purpose of 'improvement' and the dissemination of 'useful' agricultural knowledge, some of the Society's *Transactions* appeared in Bengali. It is important, then, not to assume that Indians were merely passive spectators or simply informants in relation to the Western science of the Company period. There is no doubt that in Calcutta, as in Bombay, middle-class Indians were both attracted by the practical application of Western science and technology and anxious to understand how they had given their British rulers such a remarkable ability to control and exploit the material world.[46]

---

[44] R. Scott, 'History of the Society', *Journal of the Bombay Branch of the Royal Asiatic Society: Centenary Memorial Volume* (Bombay, 1905), pp. 17, 23.

[45] Rosalind O'Hanlon, *Caste, Conflict and Ideology: Mahatma Jotirao Phule and Low Caste Protest in Nineteenth-Century Western India* (Cambridge, 1985), pp. 90–2.

[46] Jim Masselos, 'The Discourse from the Other Side: Perceptions of Science and Technology in Western India in the Nineteenth Century', paper presented at the American Asian Studies Association annual conference, Boston, 1987.

SCIENCE AND ORIENTALISM

The involvement of Indians in the scientific activities of the Company period brings us back to the question of the relationship between Western science and indigenous knowledge. In contrast to much previous scholarship, which has tended to emphasise the discontinuities that resulted from the advent of British rule, C. A. Bayly has recently argued that the 'colonial information order' was 'erected on the foundation of its Indian precursors'. Between the 1780s and 1870s, 'beyond the purview of British institutions', the agents and representatives of the older systems of knowledge survived 'traumatised, but recognisably the same'.[47] Just as Indians assimilated and adapted Western doctrines and technologies, so Western scholars and administrators embarked on a protracted intellectual engagement with Indian epistemologies that lasted well into the nineteenth century. Despite British rule, Indian systems of information and knowledge retained a high degree of autonomy and hence remained relatively inured to colonising attempts to combine political control with cultural hegemony.

Although the present work is only marginally concerned with the processes of intelligence-gathering and social communication that Bayly describes, his book also discusses other arenas of cultural engagement, such as astronomy, botany and medicine, which are more central to this discussion, and it forcefully poses the question of how closely Western science in India drew upon, or was even dependent upon, Indian ideas and information. Our historical understanding of 'colonial discourse', Bayly argues, 'must reflect the pervasiveness of Indian agency, of the Indian intellectual challenge, and of Indian cultural vitality'.[48] Although Bayly's attention is largely directed elsewhere, one place to begin considering the role and extent of the Indian contribution to the science, technology and medicine of the Company period is by looking at the institutional sites and discursive practices of colonial science itself.

Although in some respects India seemed to be located on the outer peripheries of metropolitan science, European scholars and scientists in India often saw themselves as possessing unique opportunities and responsibilities. In his opening address to the Asiatic Society in 1784, Sir William Jones laid down a challenging agenda for its members, one that gave the Society a crucial role in the scientific investigation of Asia's society, history and environment. He directed them to examine whatever was 'rare in the stupendous fabric of

---

[47] C. A. Bayly, *Empire and Information: Intelligence Gathering and Social Communication in India, 1780–1870* (Cambridge, 1996), p. 179.   [48] Ibid., p. 314.

nature', to revise the geography of Asia through 'new observations and discoveries', and to 'trace the annals and . . . traditions of those nations who, from time to time, have peopled or desolated it'. They were to illuminate the forms of government, the civil and religious institutions of the people of Asia, enquire into their astronomy, arithmetic and geometry, their systems of morality, grammar and rhetoric, their skill in medicine and surgery, their knowledge of chemistry and anatomy. The Society was to examine the agriculture, manufacturing and trade of Asia, its music, architecture and painting, even its pottery. In short, it was to reveal every aspect of 'man and nature' in Asia, 'whatever is performed by the one or produced by the other'.[49]

As Thomas Trautmann has recently argued in tracing the background to Jones's linguistic theories,[50] this wide-ranging intellectual agenda can be seen as part of an attempt to bring India and its newly discovered arts, sciences and history into a closer and more intelligible relationship with Western knowledge in a culture still dominated, despite its Enlightenment credentials, by biblical notions of the origins of race and language and of the earth's antiquity. This task was to be achieved by establishing the relationship of India's peoples and languages to those of Europe (an aim exemplified by Jones's discovery of the shared origins of Sanskrit, Greek and Latin); by comparing the traditions of the Bible with the legends of the Hindus to arrive at a mutually agreed chronology; and (one might add) by exploring the 'family' connections that linked Asia's flora and fauna, as surely as its human inhabitants, to their European kith and kin. Orientalist scholarship between the 1770s and 1840s was more, therefore, than an objective 'discovery' of India or an impartial appraisal of Eastern knowledge. In as much as one can ascribe to them a single, common agenda, the Orientalists attempted to situate India materially and culturally relative to Europe and in the process to learn from the East whatever might serve to improve or enrich their own civilisation. In common with wider Enlightenment attitudes towards non-European societies, this involved a search for correspondences and connections but it also ultimately entailed a pursuit of difference and the unfolding of a hierarchy of nations from the most primitive to the most civilised.[51] India's knowledge of itself was seldom deemed adequate or complete, but needed to be supplemented and contextualised by Western understanding. When it came to science, Jones declared in one of his more intemperate utterances, 'Asiatics' were 'mere children'.[52]

[49] Mitra, 'History', pp. 4–5.
[50] Thomas R. Trautmann, *Aryans and British India* (Berkeley, CA, 1997).
[51] John Gascoigne, *Joseph Banks and the English Enlightenment: Useful Knowledge and Polite Culture* (Cambridge, 1994), pp. 119–83.   [52] Trautmann, *Aryans*, p. 60.

Moreover, as befitted the wider Enlightenment project, Orientalism embraced 'nature' as well as 'man' and so brought a European scientific understanding of the natural world to bear on the assessment of India's material and cultural existence. It brought together (as in Prinsep's works), and without clear disciplinary distinction, philology, numismatics, ethnology, zoology, geology, and meteorology, even if, as in the Asiatic Society, language, literature and religion seemed initially to offer more fruitful avenues of enquiry into India than the natural sciences.

The underlying attitude of superiority to indigenous science can be briefly illustrated by reference to Indian astronomy, which attracted much attention in early Orientalist scholarship and which Bayly cites as an example of the contacts and continuities that existed between indigenous and Orientalist knowledge. Astronomy was one of the most technically accomplished sciences in eighteenth-century Europe, and it is not surprising that it should be well represented among the sciences of the Company period. It was also of great practical value since, until reliable timepieces and accurate surveying techniques were developed, the determination of longitude for navigation and map-making depended on observations of the moon and eclipses of Jupiter's satellites. Evidence of an ancient, still extant, astronomical tradition in India naturally aroused curiosity and suggested an area where the West might with profit learn from the East. Astronomy also raised intriguing questions about the antiquity of Indian civilisation and the extent of its technical and scientific achievements. The observatories built by Raja Jai Singh in the early eighteenth century drew appreciative comment from many European travellers. In 1777 a carefully annotated account of the observatory at Benares, reputedly dating from the age of Akbar, if not earlier, was presented to the Royal Society; its author praised the 'mathematical exactness' of the stone instruments but cast doubt on the knowledge of contemporary Brahmins.[53] Following sightings made in India in the 1760s of comets, eclipses and the transit of Venus, observatories were built between 1792 and 1842 by the Company or its Indian allies at Madras, Calcutta, Lucknow, Trivandrum and Poona. These reflected the extent of mutual interest in astronomy but, for the Company, practical needs were generally uppermost. Michael Topping, Superintendent of the Madras Observatory, thus described astronomy as 'the parent and nurse of navigation'. He hoped to see 'the charts of these eastern seas in a more correct state than those even of Europe; or at least a regular system established for the perfection of Indian geography'. His successor, John Goldingham, pursued this

---

[53] Sir Robert Barker, *An Account of the Bramins Observatory at Benares* (London, 1777).

essentially pragmatic goal by preparing two volumes of data on the satellites of Jupiter. Astronomers tried subsequently to develop a wider scientific role for their observatories but, like many other scientists in India, they found that they were expected to perform so many routine technical and administrative duties that they had little time to pursue scientific research.[54]

At first astronomy seemed a likely bridge between Western and Indian scientific knowledge. Orientalists not only noted the design and working of Indian observatories, but also, like H. T. Colebrooke, wrote sympathetically on Hindu astronomy. Europeans were employed as astronomers at Indian courts, including that of the Nawab of Awadh, who in 1831 appointed J. D. Herbert as the superintendent of his Lucknow observatory. A further stimulus to British interest in Indian astronomy was a paper by the Scots mathematician John Playfair in 1789 on 'the Astronomy of the Brahmins'. This took up the idea, earlier propounded by the French astronomers Le Gentil and Bailly, that Indian astronomy was not part of the Western astronomical tradition, traceable back to the ancient Middle East, but an entirely separate system, dating from as early as 3000 BC, and which had produced astronomical observations and predictions of astounding accuracy. Playfair's assertions in turn attracted the attention of the Asiatic Society, but, Trautmann claims, 'the question was never whether Indian astronomy had something new to offer Europe'. The aim was rather to assess the antiquity of Indian astronomy 'using the modern astronomy of Europe as the standard and means of investigation'.[55]

Although European interest in Hindu astronomy resurfaced periodically, its practical value and scientific worth were viewed with growing scepticism. A writer in the *Calcutta Review* in 1844 acknowledged that the Hindus had at a very early date been able to predict eclipses of the sun and moon with 'very considerable accuracy', but thereafter (following the standard Orientalist line) he believed that Indians had not only failed to advance further but actually 'retrograded in their knowledge of the principles of science'. Nor was that all. Far from being of great antiquity, one of the principal texts, the *Surya Siddhanta*, was now said to date only from the eleventh to thirteenth centuries AD and to be based on certain simple (and largely erroneous) principles rather than on actual observation. The author went on to declare that the 'Puranic system' of astronomy, geography and chronology was 'such a mass of absurdity and monstrous folly, that we should listen very favourably to any proposal that should promise to drive it out of the minds of the people'. It seemed to be formed

---

[54] S. M. Razaullah Ansari, 'The Establishment of Observatories and the Socio-Economic Conditions of Scientific Work in Nineteenth Century India', *IJHS*, 13 (1978), pp. 62–71; Subbarayappa, 'Science', p. 494.   [55] Trautmann, *Aryans*, pp. 79–80.

'with the special view of defying all verisimilitude, and showing the extent to which a corrupt imagination can proceed in the conception of monstrosity'. It was 'by actual measurement and actual inspection, by the measuring-rod, the theodolite and the telescope, that the Puranic idol' was to be 'demolished'.[56] As T. B. Macaulay had put it no less scathingly nine years earlier, Indian astronomy so defied credibility that it would 'move laughter in girls at an English boarding school'.[57]

Further evidence of the yawning gulf between Western science and Indian knowledge can be found in botany, a science which, following Linnaeus, had become a 'project of rationalisation, an effort to create a set of concepts and procedures that would bring uniformity, consistency, and coherence to the representation of a clearly defined domain of natural objects'.[58] As overseas expeditions left Europe 'drowning in [plant] novelties', taxonomy assumed vital importance in trying to keep track of different species and to establish a common system of plant identification. Linnaeus's binomial taxonomy and his sexual system of classification system, though the latter was later abandoned in favour of the 'natural system', suggested relatively simple means by which plants could be identified and compared worldwide.[59] Jones favoured giving Indian plants 'their true Indian appellations', using romanised Sanskrit, claiming that Linnaeus would have done likewise 'had he known the learned and ancient language of this country'. He urged botanists to study Sanskrit, to gain a better knowledge of India's medicinal plants and draw upon Indian botanical expertise.[60] But Jones's advice was largely ignored. In his *Flora Indica* Roxburgh used a motley brew of Sanskrit, Bengali, Arabic, Persian, Hindustani and Hebrew. A generation later, Buchanan recommended the use of Sanskrit place and plant names, on the grounds that these were 'known to all Hindus of learning', but he was hardly consistent in this himself.[61]

By the early nineteenth century Indian plant taxonomy was an increasingly vexed issue and reflected the wider problem of how to utilise indigenous knowledge and informants without compromising the credentials of European science. Peripatetic naturalists like Buchanan relied heavily upon information from guides and translators, but they did not want to appear over-

---

[56] 'The Astronomy of the Hindus', *CR*, 1 (1844), pp. 258, 287–8.

[57] Macaulay's 'Minute on Education' (1835), in W. T. de Bary (ed.), *Sources of Indian Tradition*, II (New York, 1968), pp. 44–6.

[58] John E. Lesch, 'Systematics and the Geometrical Spirit', in Tore Frängsmyr, J. L. Heilbron and Robin E. Rider (eds.), *The Quantifying Spirit in the 18th Century* (Berkeley, CA, 1990), p. 75.

[59] David Philip Miller, 'Introduction', in Miller and Reill, *Visions*, p. 8.

[60] Desmond, *Discovery*, pp. 53–4.

[61] D. Prain, 'A Sketch of the Life of Francis Hamilton (once Buchanan)', *Annals of the Royal Botanic Garden, Calcutta*, X (1905), p. lx.

dependent on what they took to be the highly unsystematic state of indigenous knowledge. In the course of his journey through Mysore in 1800, Buchanan noted with annoyance that villagers pretended to know the name of every plant, but actually used a limited repertoire of 'specific appellations', such as 'large', 'small', 'cultivated' and 'wild' to describe plants. Many names were thus given 'to the same species, and sometimes the same name to different species, with so little accuracy, that any person, who depends on their accounts will find himself thrown into great confusion'.[62] Robert Wight, another botanist of south India, had a similar experience, being given half a dozen different names for the same plant in the space of a few miles. He concluded that Indians had 'no means of producing an uniformity of name, and very frequently confound one name with another, so that our inserting these would only tend to mislead'. This posed a real danger when it came to identifying medicinal or poisonous plants.[63]

Although increasingly the testimony of nature was preferred to the testimony of 'natives', the rejection of local nomenclature was by no means absolute. As late as 1874 **Dietrich Brandis** urged forestry officers to learn Sanskrit and vernacular plant names, 'for in many instances they have a fixity which systematic names do not yet possess'.[64] But in general botanists favoured a tightening up of Indian taxonomy to bring it more closely into line with Europe. This was especially so when India's flora began to be studied by naturalists with no grounding in Indian languages. Hooker, who could never have assembled his vast collection of Himalayan plants without a small army of local assistants, warned against the proliferation of supposedly new species by giving indigenous names to what were in fact familiar plants. In its current 'backward state' Indian botany presented 'a perfect chaos of new names for well-known plants, and inaccurate or incomplete descriptions of new ones'.[65] In thus shedding its earlier Orientalist flirtation with Sanskrit and adhering more closely to European taxonomy, Indian plant nomenclature was indicative of a wider trend towards conformity in India's colonial science.

THE TRIGONOMETRICAL SURVEY

The Great Trigonometrical Survey of India has assumed an exemplary importance in the history of science under the Company. Susan Faye Cannon once

[62] Buchanan, *Journey*, 1, pp. 181–2.
[63] Robert Wight and G. A. Walker-Arnott, *Prodromus Florae Peninsulae Indiae Orientalis*, 1 (London, 1834), p. xxiv.
[64] Dietrich Brandis, *The Forest Flora of North-West and Central India* (London, 1874), p. xi.
[65] J. D. Hooker and Thomas Thomson, *Flora Indica*, 1 (London, 1855), pp. 36, 38.

remarked that, in seeking to establish the ideological imperatives behind British rule in India, the survey showed 'the workings of British policy better than still another study of Macaulay's education minute', adding that, as a scientific enterprise, 'it overshadowed all European geodetic operations'.[66] Surveying has served to illuminate the colonial power's ideological commitment to, and material motivation for, a sustained engagement with science on a grand (indeed subcontinental) scale, and to demonstrate the manner in which Western science in India might function more in response to local needs and conditions than at the behest of metropolitan science. With Michel Foucault in mind, the Indian surveys (and the accompanying exercises in map-making and topographical depiction) have been seen as constituting 'grids' of disciplinary knowledge, a means by which the British established a comprehensive network of surveillance and control over the Indian countryside. Matthew Edney has recently proposed (primarily in order to dispute that such a totalising project was ever in practice realisable) the idea of survey as a kind of scientific 'panopticon' designed to give the imperial power the capacity to 'see' and 'know' India in a systematic and scientific manner. Although he concludes, predictably, that the survey was a 'flawed geographic panopticon', we are none the less presented with the idea of the trigonometrical survey as standing at the forefront of British efforts to colonise India through science and technology and of scholars' attempts to establish the wider significance of colonial scientific and technological practices.[67]

Initially, surveying in India was closely bound up with a quest for dependable maps. European mapping of South Asia began in earnest with the French in the early and mid eighteenth century but advanced more rapidly after the English East India Company acquired Bengal and parts of southern India. Early map-making was primarily carried out through detailed, descriptive accounts of the features and topography of specific localities, as in the case of Mackenzie's survey of Mysore, or through military surveys taken along a line of march. These in turn provided the basis for regional maps of the kind that established James Rennell's cartographic reputation. Appointed India's first Surveyor-General in 1767, Rennell conducted a topographic survey of Bengal and Bihar, subsequently published as his *Bengal Atlas* of 1780, and in 1783 produced a *Map of Hindoostan*. But these techniques of map-mapping were notoriously haphazard, leaving blank large areas of the map away from the main routes and towns and even the distances between known points little more

---

[66] Susan Faye Cannon, *Science in Culture: The Early Victorian Period* (New York, 1978), p. 251.
[67] Edney, *Mapping*, pp. 24–5, 53–4, 325.

than conjecture. Astronomical observations provided a partial corrective, and were used by Topping and Goldingham at the Madras Observatory to calculate distances along the Coromandel coast, but this technique, too, was prone to error and difficult to use as a basis for systematic mapping over a wide area. More effective survey and mapping techniques were at this time being developed in Europe, of which triangulation was the most important. The British Trigonometrical Survey, dating from 1784, and the Irish Ordnance Survey provided important technical and organisational precedents for the trigonometrical survey and map-making in India; to this extent metropolitan precedents were clearly of great significance.[68]

Beginning in 1799, **William Lambton**, a British infantry officer, employed the new technique of triangulation, measuring out baselines with steel chains 100 feet long, to determine the distance between Madras and the west coast. Checked against astronomical readings, the survey, completed only in 1817, could then be used as a reliable basis for measuring and mapping south India. The perceived success of this technique prompted the founding of the Great Trigonometrical Survey in 1818, with Lambton in charge of its India-wide operations. Its task was to extend triangulation from the southern peninsula into central and northern India (through areas but recently wrested from Maratha control), thus providing a basis for mapping the entire country. By the time of Lambton's death in 1823 the survey had been carried as far as Berar in central India. Apart from the appeal to the state of its practical utility, the survey was seen by men like Lambton as being of immense scientific significance: British rule in India created a unique opportunity to measure a larger segment of the earth's surface than had hitherto been attempted anywhere else. The precise dimensions of the globe could thus be determined with unprecedented accuracy.

Under George Everest, Lambton's successor, the baselines were painstakingly re-measured with new equipment from Britain (the old measuring chains being replaced by 'Colby's compensation bars' to minimise variation in the length of the measuring instruments) and the line of triangulation was pushed northwards to the Himalayas. By the time Everest retired in 1843 a 'great meridional arc' had been measured by triangulation over 1,500 miles from Cape Comorin in the south to Dehra Dun at the foot of the Himalayas. With supplementary surveys (the Calcutta, Bombay and Karachi longitudinal series) linking eastern and western India, and with extensions into the northwest and Burma, the survey should, in theory, have provided an accurate basis by the

[68] Ibid., pp. 244–8, 288.

1860s for mapping the entire subcontinent. In technique, as in scale, the Indian survey was in advance of any geodetic survey undertaken at the time either in Europe or by other European powers overseas. Despite its British and Irish antecedents, and despite the extensive reliance on imported equipment, it reflected the initiative and determination of the Company's own military and scientific servants. It showed that colonial India was capable of generating its own techniques and institutions and was more than a mere camp-follower of metropolitan science. Despite Lambton's initial lack of support from the Madras government, it also demonstrated that the Company was capable (in this instance at least) of a sustained commitment to science, to the extent of making the survey in effect India's first dedicated scientific service.

Celebrated in science, the survey also served more immediately utilitarian ends. Surveying by land was a necessary adjunct to measures to determine the precise location of India's ports and coastal features and thus ensure greater security for the navy and maritime trade. As in Britain and France at the time, surveying operations reflected the close collaboration between cartography and the army, with its urgent need for maps that would reliably serve its needs in wartime. The Indian survey significantly originated under the auspices of an expansionist colonial state at a time of widespread warfare and political uncertainty in south and central India, and drew on the precedents of military engineers in Britain such as William Roy.[69] Both Mackenzie's topographical survey of Mysore and Lambton's trigonometrical survey of the peninsula arose directly out of the defeat of Tipu Sultan in 1799. Reliable maps were needed for military use and, even after the defeat of the Marathas in 1818, continuing resistance in Punjab and the perceived threat of Russian expansionism gave surveying a lasting military significance. Looked at in broader terms, the survey was a means of addressing what one member of the Governor-General's Council called the problem of 'the immense regions subject to . . . British authority or influence', in other words of how to transform India, with all its political and topographical diversity, into a single, mapped entity, an ordered and rationalised imperial space.[70]

And yet, for all its military and political importance in 'constructing' and 'disciplining' India, the survey, Edney has argued, never fulfilled its grand objectives. The 'cartographic anarchy' of the late eighteenth and early nineteenth centuries was only partly resolved by triangulation. The survey began

[69] Sven Widmalm, 'Accuracy, Rhetoric, and Technology: The Paris–Greenwich Triangulation, 1784–88', in Frängsmyr et al. (eds.), *Quantifying Spirit*, pp. 201–3.

[70] Matthew Henry Edney, 'Mapping and Empire: British Trigonometrical Surveys in India and the European Concept of Systematic Survey, 1799–1843', Ph.D. thesis, University of Wisconsin, Madison, 1990, p. 240.

only after much cartographic labour had already been expended, and its loosely constructed, Meccano-like skeleton was never sufficiently comprehensive to provide a basis for mapping the whole of India. Inconsistencies in Company policy, intra-provincial rivalries, the smallness of the survey establishment, the vulnerability of survey parties to ill health and accidents, the enormous cost and physical difficulty of surveying across India's rivers, mountains, deserts and jungles – all rendered a completely scientific survey an unrealisable goal. Edney sees this as calling into question the whole idea of the survey as a kind of omniscient, Foucauldian vision of empire, concluding that the 'apparent perfection of the geographical panopticon promised by the Great Trigonometrical Survey' was in the end 'an empiricist delusion'.[71] Even so, even if triangulation failed to deliver a comprehensive survey, it is striking how, under an often lackadaisical, cash-conscious Company, a vast scientific project was effectively sustained over several decades and with a conscious acknowledgement that it might serve ideological as much as practical objectives. Few other scientific enterprises, under Company or Crown, were ever so consistently favoured and supported.

There is a further point to be stressed about the survey. As with many other areas of early colonial science, it has rightly begun to be appreciated that mapping and surveying were heavily reliant on Indian skill and agency for their execution. Particular significance has attached to the career of Radhanath Sikdar, who worked for the survey from 1831 onwards, and the 'pandits' who after 1860 secretly carried surveying deep into Tibet, well beyond the reach of British surveyors.[72] But as Edney demonstrates, the technical and organisational precedents for the survey lay, not in India, but in the cadastral and topographical surveys developed in Europe since medieval times. Moreover, despite the genuinely appreciative remarks made by Europeans about individual survey workers such as Sikdar, within the survey as a whole Indians were strictly confined to subordinate posts. This was indicative of the manner in which such scientific endeavours and technological achievements were seen to reflect the superiority of Europe itself. 'By measuring the land,' Edney remarks, 'by imposing European science and rationality on the Indian landscape, the British distinguished themselves from the Indians: they did science, the Indians did not.' In form and function the survey articulated a British mastery over the Indian landscape, reducing the 'mystical, religious, Hindu space of India to a rational, scientific, imperial structure of space'.[73]

[71] Edney, *Mapping*, p. 30.
[72] Kumar, *Science*, p. 59; Kapil Raj, 'Les hommes – instruments de sa majesté', *Recherche*, no. 298 (1997), pp. 74–80.   [73] Edney, 'Mapping', p. 19; *Mapping*, p. 32.

GEOLOGY

Although natural history and environmental sciences were among the main disciplinary sites of the Company period, some sciences found more favour than others. Despite significant descriptive accounts of snakes, fishes and birds, and pioneering work by Buchanan, Hodgson and Blyth, zoology was slow to establish itself as a distinct discipline in India, despite the wealth of material available. One reason for this was the lack of official interest and of an institutional base. Less obviously a 'useful' science than trigonometry or botany, zoology continued to reflect the passions of the amateur enthusiast, collector and hunter. T. C. Jerdon's *Mammals of India* in 1874 marked a move towards more a systematic study but, rather than invoke the material concerns of state and society, his manual was primarily intended for 'sportsmen and observers'.[74] *Shikar* (hunting) long remained one of Indian zoology's principal sources of inspiration and information.

Geology, by contrast, progressed rather more rapidly from amateurism to professional status and state support. Initially, as Andrew Grout has pointed out, geological investigation in India was slow to develop. One reason for this 'antipathetic attitude' was concern by the Company and influential pressure groups in London that India should not compete with Britain's own production of copper and other minerals. Another factor was a presumption that India was, and should remain, primarily a site for agricultural production, but a general ignorance of India's resources, an apparent lack of technical expertise and an exaggerated sense of the technological difficulties involved further discouraged systematic investigation before the 1830s.[75] Early accounts of mineralogy and stratigraphy were accordingly descriptive and sporadic. Reflecting on this in 1833, James Calder hoped that, just as Indian botany had 'found its Linnaeus' in Roxburgh, so 'we may yet see the treasures of the animal and mineral kingdoms unfolded to us by a Humboldt and a Cuvier'.[76] There was an inhibiting feeling, too, that India's geology had less to offer of scientific interest than its botany or zoology. As late as 1846 Captain Newbold extended to India Humboldt's observation that, whereas a host of new plant and animal species greeted visitors to the Americas, 'in the rocks we still recognise our old acquaintances; the same granite, the same gneiss, the same micaceous schists, quartz rocks, etc.'[77] The vast plains and deep alluvium of north India and

[74] T. C. Jerdon, *The Mammals of India* (London, 1874), p. v.
[75] Andrew Grout, 'Geology and India, 1775–1805: An Episode in Colonial Science', *SAR*, 10 (1990), pp. 1–18; Grout, 'Geology and India, 1770–1851'.
[76] James Calder, 'General Observations on the Geology of India', *AR*, 18 (1833), p. 1.
[77] T. J. Newbold, 'Summary of the Geology of Southern India', *JRAS*, 8 (1846), p. 156.

Bengal at first stirred little geological excitement and yielded few fossils. In remoter, less populated regions, 'impenetrable jungles' allowed only 'a vague and scanty knowledge' of underlying strata; wild animals and malaria added to the hazards of geological reconnaissance, as they did to those of surveying. Tigers, having 'no regard for science', would 'eat up a geologist as soon as they would any other animal'. Until canals and railways ran a scalpel through the landscape, there were few deep excavations to reveal rock strata and expose the fossil record. Compared with Europe, the geology of India appeared 'far less complex', even 'monotonous'.[78] Gradually, though, the distinctive features, the scientific opportunities and the practical rewards of Indian geology became apparent. The nature and origins of *kankar* (the calcareous nodules found extensively in north Indian soils) and laterite (first described by Buchanan in Malabar in 1800), the extraordinary depth of the alluvial deposits beneath Calcutta (revealed by boring for water to a depth of 500 feet), the vast extent and thickness of the Deccan lava flows (a feature, according to Newbold, 'not to be surpassed in any other portion of the globe'), the momentous discovery of the Siwalik fossils, and (in an age awakened to the 'sublime' quality of mountain scenery) the fascination of the 'stupendous' Himalayas had earned for Indian geology by mid-century international interest.[79] It was to this science, above all others, Larwood grudgingly conceded, that 'Indian fieldworkers contributed most significantly'.[80]

If, initially, geology reflected the individual enthusiasm of soldiers and surgeons rather than any official policy of mineral extraction, by the 1830s the Company was becoming more aware of geology's economic utility and newfound scientific standing. The arrival of steamships, with the prospect of railways to follow, made a search for coal imperative, and in 1836 a Committee for Investigating the Coal and Mineral Resources of India was duly created. The Industrial Revolution had formed a powerful image in British minds of the transforming power of coal and iron; coal was expected in India similarly to hasten the spread of 'European arts and improvements'. 'The history of coal', declared M'Clelland as Secretary to the Coal Committee, was 'sufficient to show how the manufactures of a nation may be improved by its judicious application, and there is no reason why its beneficial effect should be here less salutary than elsewhere'.[81] The committee sought to establish the location,

---

[78] Calder, 'Observations', pp. 6, 20; 'Geology of India', *CR*, 9 (1848), p. 327; R. Baird Smith, 'Notes Illustrative of the Geology of Southern India', *CJNH*, 1 (1840), p. 188.

[79] Buchanan, *Journey*, 11, pp. 440–41; Calder, 'Observations', pp. 2–4, 17; R. Baird Smith, 'On the Structure of the Delta of the Ganges, as Exhibited by Boring Operations in Fort William, AD 1836–40', *CJNH*, 1 (1840), pp. 324–43; Newbold, 'Summary' (1846), p. 144.   [80] Larwood, 'Science', p. 71.

[81] *Reports of a Committee for Investigating the Coal and Mineral Resources of India* (Calcutta, 1838), pp. 16–17.

extent and accessibility of Indian coalfields, but its slow progress and lack of practical results illustrated the dilatory manner in which even such seemingly vital investigations often proceeded under the Company. 'The government move slowly in matters of improvement unconnected with the ordinary duty of government in judicial, revenue and political departments', a disheartened M'Clelland wrote to Lyell in 1844. 'Reforms in matters involving scientific questions such as the examination of the coal formations of India . . . stand every chance of being laid aside until forced upon its attention.'[82] In 1846, D. H. Williams of the British Geological Survey arrived to oversee the exploration of India's coal resources, but his death in 1848 brought further delay. The Court of Directors then sent out Thomas Oldham, former Director of the Geological Survey of Ireland, whose arrival in 1851 marked the birth of a more professional geological agency and, in the wake of coal, the beginning of a more systematic investigation of India's stratigraphy and mineralogy. Over the next fifteen years, as India embarked on its railway age, twenty-seven coalfields were investigated. Under the impetus of economic and technological change, geology shed its 'amateur' status to become an integral branch of the colonial administration. In this respect, the Geological Survey of India, like the Trigonometrical Survey, represented the emerging role of scientific and technological agencies in colonial state formation and in defining the expanding roles and responsibilities of the modern state in its Indian setting.

BOTANY

Utility was likewise the hallmark of colonial botany, but (like the Trigonometrical Survey) its significance extended beyond narrow instrumentality. Like geology, but from a much earlier date, Indian botany was seen both as having a practical utility and as being of interest to the international scientific community. A sound knowledge of Indian plants and their properties was an important asset to a Company that traded in spices, indigo and other natural products and whose European servants were constantly plagued with sickness. In 1778 the Danish surgeon-botanist J. G. Koenig, previously attached to the Tranquebar Mission, became the first naturalist formally employed by the East India Company. Though pioneering work on Indian flora had been carried out by the Portuguese physician Garcia d'Orta in Goa in the sixteenth century and by the Dutchman Heinrich Van Rheede in Malabar in the seventeenth, it was Koenig who introduced Linnaean taxonomy to India

---

[82] Sangwan, 'Reordering', p. 296.

and linked the study of Indian flora to the emerging global networks of botanical science. His work impressed the entrepreneurial Banks in London and paved the way for Roxburgh's research in south India.[83] Of all the sciences of Company India, botany was surely the least 'provincial' and the most closely embroiled with metropolitan science.

In Europe's 'age of improvement' botany was essential for the introduction of new plants to augment revenues from agriculture and trade and to protect populations from disease and famine. In India such objectives, for all their underlying commercial self-interest, lent themselves to representations of foreign rule in humanitarian guise and a judicious blend of profit and philanthropy was evident in the early history of India's botanic gardens. The first of these, at Calcutta, was established following a proposal in 1786 by Colonel Robert Kyd. Drawn more to utilitarian horticulture than to botanical science, Kyd urged that a garden be established for practical purposes and not for collecting rare plants 'as things of mere curiosity or furnishing articles for the gratification of luxury'. The introduction of exotics, like sago and date palms, would, he argued, supplement India's apparently meagre store of food plants and protect against the famines that had several times in recent decades devastated Bengal. (At about the same time in south India famine encouraged Roxburgh at Samalkot to think along similar lines: botany thus gave early warning of the importance famine was to hold for the entire development of science, technology and medicine in nineteenth-century India.) Kyd further suggested that, at a time of continuing conflict with Holland and France, the British would benefit commercially from having their own supply of cinnamon, still a monopoly of Dutch Ceylon, as well as indigo, cotton, tobacco, coffee and tea. The successful introduction and dissemination of such plants in Bengal would bring profit to the Company and enable the British to 'outstrip our rivals in every valuable production which nature has confined to this part of the globe'. The Court of Directors was sufficiently impressed by Kyd's economic arguments and the enthusiastic support given by Banks, an influential advocate for the contribution botany could make to trade and 'improvement', to sanction the proposal in 1787.[84]

With Kyd as its Honorary Superintendent, a horticultural garden was laid out at Sibpur, across the Hooghly from Calcutta, but by the time he died in 1793 the scheme had begun to flounder. The garden received a large number of plants from Southeast Asia and elsewhere, but many of the most important

[83] I. H. Burkhill, *Chapters on the History of Botany in India* (Delhi, 1965), pp. 4–20.
[84] Kalipada Biswas, *The Original Correspondence of Sir Joseph Banks Relating to the Foundation of the Royal Botanic Gardens* (Calcutta, 1950), pp. 3–12; Gascoigne, *Banks*, pp. 185–236.

introductions, including tea bushes from China, failed to thrive in Calcutta's heat and humidity. Under Roxburgh, Kyd's successor, the utilitarian engagement with economic botany was retained and extended (as through Roxburgh's investigation of vegetable fibres), but the gardens became more a centre for plant collection and classification than Kyd had intended and the practical improvement of Indian agriculture remained a relatively neglected item on the agenda of Company science. Roxburgh confined himself to a 'quiet life in the Calcutta garden', where he classified over 2,000 species, leaving to Buchanan and Wallich the 'restless life' of plant collection in Nepal, Assam and Burma, before they in turn took charge of the gardens after Roxburgh's departure in 1813.[85]

Though much the most famous, Calcutta was not India's only botanic garden. By the 1850s there were several such gardens distributed throughout India; one of the most important of these was established at Saharanpur, on the southern fringes of the Himalayas. In 1817 Surgeon George Govan proposed Saharanpur as a suitable location for growing plants that would not thrive in Calcutta.[86] Govan left in 1821, but over the following decades the 40 acre garden flourished and the sale of plants from Saharanpur, including many of the trees that lined the banks of the north Indian canal system, yielded a modest income. As the garden's Superintendent from 1823 to 1831, J. Forbes Royle, investigated the properties and commercial potential of cotton, tobacco and other 'useful plants'; tea was successfully introduced, but cinchona failed. He tried to impress on the Company the vast range of India's vegetable products and their value alike for 'the comforts of the people' and 'the wants of a great empire'.[87] Royle, who continued to stress the commercial value of India's economic botany after retiring to London, was (like many other surgeon-naturalists) well aware how much science depended on the Company's favour, but also how sluggish it could be even where agricultural improvement was concerned.

Botany also illustrates some of colonial science's other roles. In the 1790s the Baptist missionary William Carey laid out a 5 acre garden and herbarium at Serampore, north of Calcutta, in which he tried to grow, among other plants, the hedgerow flowers that reminded him of his native Northamptonshire. There was more to this than homesick sentimentality. Botany, for Carey, as for many Christian naturalists of his day, was a celebration of God's handiwork,

---

[85] Burkhill, *Chapters*, p. 28.

[86] H. Montgomery Hyde, 'Dr George Govan and the Saharanpur Botanical Gardens', *JRAS*, 49 (1962), pp. 47–57.

[87] J. Forbes Royle, *Illustrations of the Botany and Other Branches of the Natural History of the Himalayan Mountains and of the Flora of Cashmere*, 1 (London, 1839), p. 442.

and when he edited his friend Roxburgh's *Flora Indica* for publication, the title page duly bore the inscription: 'All Thy Works Praise Thee, O Lord.' Carey, like Kyd, also valued the practical, 'improving' side of botany and he was largely instrumental in setting up the Agricultural Society of Calcutta in 1820.[88] There is no doubt, too, that Company servants derived aesthetic pleasure from the study of Indian flora and from the coloured plates and drawings that adorned their botanical works. Botany was an art form as much as a science, especially at the hands of illustrators like Vishnu Prasad and Laksman Singh, whose paintings for Hodgson, Wallich and Royle were both a necessary means of preserving and disseminating plant data and a means of utilising the decorative skills of Indian artists.

A further aspect of botany was its intimate relationship with medicine. The science of botany had emerged in medieval and early modern Europe hand-in-hand with medicine; the identification of medicinal plants and a knowledge of their properties long remained an essential part of medical training. Although in Europe botany had achieved a degree of intellectual independence from medicine by the mid-eighteenth century, it still formed an integral part of the education of Company surgeons, especially those trained at Edinburgh University – Roxburgh, Buchanan and Falconer among them. Of the forty-two individuals I. H. Burkhill identified as practising botanists in the period up to 1840, twenty-eight were surgeons, as were roughly a quarter of those active in the second half of the century.[89] That surgeons enjoyed a comparable ascendancy in zoology, and to an almost equal extent in early Indian geology and forestry, made the Company's medical service, in effect, an all-purpose scientific agency and for a long time obviated the need for other, more specialist, bodies. Not until 1890 did India have a Botanical Survey, or till 1916 an equivalent organisation for zoology.

In India the close association between botany and medicine was strengthened by the cost-driven search for local substitutes for imported drugs and the need to know the names and properties of plant medicines used by Indian physicians.[90] Like astronomy, this could constitute a significant link with indigenous knowledge, but it could be a critical as well as appreciative engagement. At Saharanpur, Royle, responsible for the medical needs of a large military station as well as for the botanic garden, investigated items of materia medica sold in local bazaars and speculated on the relationship between Indian medicine and that of the Middle East and Europe. Ultimately, he

---

[88] F. Deaville Walker, *William Carey: Missionary Pioneer and Statesman* (London, 1926), pp. 302–5; O. P. Jaggi, *Science in Modern India* (Delhi, 1984), pp. 38–9.   [89] Burkhill, *Chapters*, pp. 27, 272.
[90] John Fleming, 'A Catalogue of Indian Medicinal Plants and Drugs', *AR*, 11 (1810), pp. 153–96.

reversed his original assumptions about the Graeco-Arabian origins of Indian pharmacology and, in a seminal piece of Orientalist reasoning, concluded that the Indian medical system was much older.[91] In pursuit of his botanical and medical interests, Royle enlisted the help of merchants who travelled to Kashmir to bring back seeds and plants for him, some of which he successfully propagated at Saharanpur. Many of the plants investigated were of medicinal value and Royle sought to determine which plants Indian physicians used and to describe scientifically the nature and properties of each species; the results were published in his study of Himalayan flora in 1839. Since Hindu pharmacology was 'noted for the employment of powerful drugs' – such as aconite, nux vomica and croton – it was essential to know which plant was which, which parts of a plant should be used and for which purpose, and to standardise the taxonomy of medicinal plants across India as a whole.[92] On leaving India, Royle returned to London where he was Professor of Materia Medica at King's College from 1837 to 1856. Like Falconer, Royle was one of several Company surgeons who subsequently established scientific or academic careers in Britain, and who, while maintaining an interest in India, made significant contributions to metropolitan science.

SCIENCE IN THE TROPICS

At a time when Orientalist scholarship was examining the connections between the languages, literatures and religions of India and Europe, the sciences of the period were engaged in the parallel exercise of situating India relative to the rest of the natural world. In part this was an exploration of universality, fitting information about Indian stratigraphy or ornithology into an emerging global picture of the fossil record and geological epochs or of avian species and their distribution. At the same time, however, and with growing force, the scientists of the Company period were turning away from indigenous knowledge and addressing themselves directly to the forms and properties of India's natural history and physical environment. In so doing they were investigating not only those aspects of nature that linked India to the more familiar scientific domain of Europe but also those that appeared to differentiate India, as an essentially 'tropical' country, from more temperate lands. This investigation established tropicality as one of the central paradigms of colonial science in India (and India, equally, as one of the primary sites for the investigation of tropical plants, animals and diseases). At the same time it

---

[91] J. F. Royle, *An Essay on the Antiquity of Hindoo Medicine* (London, 1837).
[92] Royle, *Illustrations*, I, pp. 47–8.

gave emphasis to the external, judgemental nature of colonial science and the manner in which science informed colonial attitudes to Indian society and culture.

The Western identification of tropical, as opposed to temperate, regions of the globe has a long history. It can be traced back to the early European voyages of exploration, but it was strengthened in the seventeenth and eighteenth centuries by the apparent failure of white settlement in the West Indies and the lowlands of the tropical American mainland. The conviction grew that such regions were unsuited by climate and disease for European settlement and agriculture and better fitted for African slavery and plantation crops. By the mid-eighteenth century, as European rivalries intensified, an extensive literature in natural history, in topography and, above all, in medicine emerged, depicting the West Indies as a hazardous environment for peoples from the temperate lands of Europe. From this cardinal site, ideas of tropical 'otherness' circulated around the globe. At first India did not seem to fit readily within this emerging temperate/tropical paradigm. From at least the time of the famine in Bengal in 1770, with its estimated 10 million deaths, India was seen as a land of recurrent dearth and disease, an image that was sustained and intensified by the widespread famines and epidemic mortality of the nineteenth century, especially after 1860.[93] India's droughts and famines seemed to contradict the general perception of tropical fertility and natural abundance that, Malabar and Assam apart, seemed more appropriate for Sri Lanka and the Malayan archipelago than for India. Nevertheless, by the 1820s and 1830s the word 'tropical' was coming into increasingly common use in botanical, medical and even geological texts in India as a more positive evaluation of Indian civilisation began to recede and to be replaced by a heightened sense of Indian vulnerability to an often cruel and capricious world of nature.[94] The creation of India's botanic gardens and their connection with famine is indicative of the extent to which the population of India appeared unable to sustain, unaided, adequate levels of subsistence: hence the perceived need for European scientific knowledge and technological expertise to redeem them from their servitude.

These negative representations of India mingled, somewhat paradoxically, with growing commercial and botanical interest in the value to Europe of India's tropical agriculture. At about the time Kyd was proposing a botanic

---

[93] For the principal famines of the period, see Leela Visaria and Pravin Visaria in Dharma Kumar (ed.), *The Cambridge Economic History of India, II, c.1757– c.1970* (Cambridge, 1983), pp. 528–31.
[94] David Arnold, 'India's Place in the Tropical World, 1770–1930', *Journal of Imperial and Commonwealth History*, 26 (1998), pp. 1–21.

garden for Calcutta, Banks, as President of the Royal Society and scientific adviser to the East India Company in London, was urging the Court of Directors to develop its newly acquired Indian territories as a tropical dependency. Banks proposed the introduction or expansion of crops such as sugar, cotton, coffee, cochineal and indigo, which could not be grown in Europe but were of great value and importance to British trade and manufacturing. The size of India, the diversity of its products, the variety of its climates and soils, the abundance of its labour – all made India appear an ideal but under-utilised estate that could serve and complement British agriculture and industry. In this role, Banks believed, India would soon outstrip the West Indies.[95] In 1788 he presciently observed that the country between Bengal and Bhutan might be ideally suited for the cultivation of tea. This was a region, he added in one of the first references to India's tropicality, where 'in a few days you get from the tropical heats and consequently tropical productions to a climate similar to that of Europe'.[96]

Along with medicine, botany was one of the principal channels through which the concept of a tropical India became established in colonial discourse. As early as 1829 the Agricultural Society of Calcutta saw one of its principal aims as being to introduce plants 'from other tropical countries, especially America and the West Indies'.[97] At the Saharanpur garden, Royle took up a similar task. Like many of his contemporaries, he greatly admired the work of the German naturalist Alexander von Humboldt, and made frequent reference to 'the illustrious Humboldt' in his account of Himalayan flora. In fact, Royle was in many respects a poor representative of 'Humboldtian science'.[98] He lacked the German's pioneering use of scientific instruments, measurement and mapping, and could not match his Romantic enthusiasm for the aesthetic delights and spiritual experience of landscape and vegetation. He was also less of a traveller: most of Royle's knowledge of the Himalayan region was remarkably secondhand. But Humboldt's descriptions of South America enthused Royle to see an analogy between the effects of elevation and aspect on the climate and botany of the Andes and those of the Himalayas, and he developed the idea of India having tropical, temperate and alpine zones as a matter of great practical significance. If Calcutta's botanic garden gave institutional form to the India of the tropics and illustrated the kinds of plants that might

---

[95] David Mackay, *In the Wake of Cook: Exploration, Science and Empire, 1780–1801* (London, 1985), pp. 174–6. [96] Harold B. Carter, *Sir Joseph Banks, 1743–1820* (London, 1988), p. 272.
[97] *Gleanings*, 1 (1829), p. 208.
[98] Cannon, *Science*, pp. 73–110; Malcolm Nicolson, 'Alexander von Humboldt and the Geography of Vegetation', in Andrew Cunningham and Nicholas Jardine (eds.), *Romanticism and the Sciences* (Cambridge, 1990), pp. 169–85.

be imported and naturalised from tropical America, Africa and Southeast Asia, so, at 1,000 feet above sea level and lying at the foot of the Himalayas, Saharanpur represented a temperate environment. Some tropical plants might be grown there, but the higher one moved into the foothills and valleys of the Himalayas the more one passed, as in the Andes, from tropical to temperate, and ultimately to alpine, vegetation. There was accordingly hardly any 'useful' plant – tropical or temperate – that could not be grown somewhere in India.

The growing perceptual division between tropical and temperate South Asia was accentuated by the fact that early nineteenth-century botanists were interested in the exuberance and diversity of tropical plant life – as in northeastern India and Malabar – or in the temperate species of the Himalayas and Nilgiris, but found little of interest in the cultivated plains in between. For Royle, as for many of his contemporaries, the Himalayas held a special fascination that elided science with sentiment. Just as the mountain range appeared as a bridge between the languages and races of southern Asia and Europe, so was it the magical trysting place where the fauna and flora of Europe and western Asia joined company with those of Southeast Asia and China. After the alien, torrid plains, the Himalayas represented to naturalists, as to many other Europeans, a kind of spiritual homecoming. Royle explained that, on arriving 'in a tropical country' like India, a European was first struck by 'the magnificent peculiarities of its vegetation'. But after an extended residence these pleasures palled; his attention was 'more quickly excited by the re-appearance of forms with which he was familiar in his youth, and which characterise the more humble and verdant, but not less beautiful flora of temperate climates'. The higher one climbed in the Himalayas, the more 'the plants of India disappear, and we are delighted at finding the increase in number and variety of those belonging to the European genera'. At first there were 'only a few struggling towards the plains, which in a more temperate climate would be their favourite resort', and it was 'not until we have attained a considerable elevation that, having apparently lost all traces of tropical vegetation, we enter a forest of pines or oaks, and lofty rhododendrons, where none but European forms are recognisable'.[99]

The scientific representation of India in terms of a torrid core and temperate rim received authoritative endorsement from Hooker in the 1840s. Following his earlier travels in Antarctica and New Zealand, Hooker was eager to acquire a corresponding knowledge of the tropics. India initially disappointed him: the vegetation of Bengal was not tropical enough for his Humboldtian fervour, and only when he travelled up into the Himalayan

---

[99] Royle, *Illustrations*, I, p. 15.

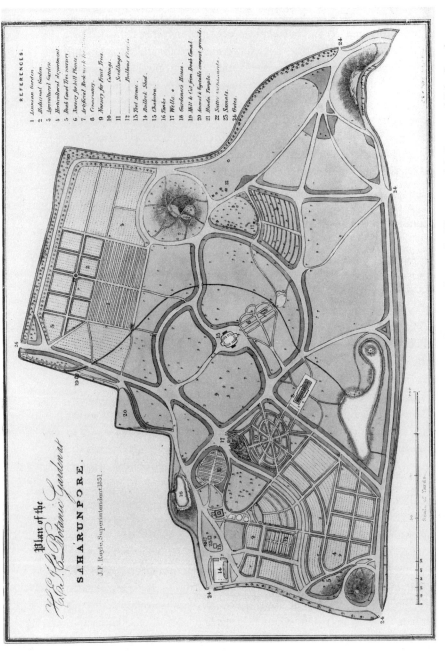

Plan of the H.E.I.C.'s Botanic Garden, Saharunpore, 1831, from J. Forbes Royle, *Illustrations of the Botany and Other Branches of the Natural History of the Himalayan Mountains and of the Flora of Cashmere*, II, London, 1839, map between plates 10 and 11; Wellcome Institute Library, London

foothills did he begin to find the kind of 'tropical luxuriance' he had anticipated, but he was even more enthused by the temperate flora that lay higher up. The violets, strawberries, oaks and birches 'vividly' recalled England: some flowers were 'so notoriously the harbingers of a European spring that their presence carries one home at once'. Hooker declared Darjeeling an ideal site for a sanatorium, a place where European children, brought up 'sickly, pallid or yellow' from the plains, were swiftly transformed into 'models of rude health and activity'.[100] In what might be taken as a precedent for the subsequent location of scientific research institutes in Indian hill-stations, Hooker then settled down with his host, Hodgson, to study India's natural history from his Himalayan eyrie.

Orientalist scholarship and a tropical/temperate paradigm of environment and natural history were not necessarily irreconcilable pursuits; in many ways they formed part of the same exploratory, utilitarian, integrationist agenda. Jones, the Orientalist, and Banks, the economic botanist, shared much in common in their attitudes to empire and 'improvement'. Without much grounding in Sanskrit, Royle could produce a seminal text establishing the antiquity of 'Hindoo medicine'; he could also, as surgeon-botanist, extol the virtues of developing India's temperate as well as tropical agriculture. With its *sati* monuments, Chinese tea bushes and Linnaean garden, laid out on the site of what had once been a Rohilla chief's pleasure garden, Saharanpur was itself suggestive of the eclecticism of Company science. But, increasingly, the pursuit of botany, like geology and zoology, led science to depart further and further from Sanskrit texts and Brahmin pandits, and brought it instead into closer association with the land forms, natural history and human inhabitants of the remoter parts of India. While the Company's own relationship with science remained, at best, equivocal, Western science, as part of the cultural world-view of the colonial elite, increasingly framed the British understanding of India, juxtaposing, as in the case of Awadh on the eve of its annexation, the enticing potentialities of vast natural resources with the glaring deficiencies of

---

[100] Joseph Dalton Hooker, *Himalayan Journals*, 1 (London, 1854), pp. 109, 120.

1 Linnean Garden.	9 Nursery for Fruit Trees.	17 Wells.
2 Medicinal Garden.	10 .... Cuttings.	18 Gardener's House.
3 Agricultural Garden.	11 .... Seedlings.	19 Mill & Cut from Doab Canal.
4 Horticultural department.	12 .... Bulbous Plants.	20 Animal and vegetable compost grounds.
5 Doab Canal Tree nursery.	13 Tool House.	21 Hindu Temple.
6 Nursery for hill Plants.	14 Bullock Shed.	22 Suttee monuments.
7 Artificial Rock work for Plants.	15 Chabutra.	23 Samats.
8 Conservatory.	16 Tanks.	24 Gates.

its weak and decadent rulers.[101] Where Jones and the early Orientalists had seen India as an ancient, if decayed, civilisation, worthy of comparison with Greece and Rome, scientists of a later generation turned instead to independent observation of the natural world, or pondered the tyranny with which nature ruled the lives of races bereft of modern science and technology. It was increasingly axiomatic that without them there could be no durable civilisation, no lasting 'improvement', in the tropics.

[101] 'The Physical Capabilities of Oude', *CR*, 24 (1856), pp. 415–44.

CHAPTER 3

WESTERN MEDICINE IN AN INDIAN ENVIRONMENT

Medicine occupied a central place in Western scientific thought and activity in nineteenth-century India. There were many reasons for this. Firstly, the Colonial Medical Service was one of the principal scientific agencies in India during the Company period and for several decades thereafter. Company surgeons and their successors under the Crown provided a large share of the botanists, geologists, zoologists, meteorologists, foresters and other specialists. Secondly, partly because of their wide-ranging scientific brief, medical personnel had a vital role in the European investigation of the Indian environment (including its topography, climate and diseases), and hence in understanding how nature fashioned the human condition in India. Thirdly, to a degree unparalleled in other scientific fields and matched by few aspects of technological change, medicine represented direct intervention in, and interaction with, the social, cultural and material lives of the Indian people. This dual engagement – with the environment and with culture – helped fashion not only the distinctive character and preoccupations of India's colonial medicine, but also the manner of its Indian reception and assimilation. Although medical and sanitary intervention was initially driven by the scientific interests of the colonial state, over the course of the century medicine began to serve other agendas and to inform a wider cultural and political dialogue.

From the perspective of medical history, the demise of the East India Company in 1858 was not in itself particularly momentous, and it is more appropriate to see the nineteenth century as a whole. Some developments can usefully be traced through to the 1910s and 1920s, but this chapter is mainly concerned with the period up to the mid-1890s, when, in the wake of Robert Koch's identification of the cholera bacillus and **Ronald Ross**'s discovery of the mode of malaria transmission and the outbreak of bubonic plague in India, there was a shift away from the environmental paradigm that had dominated nineteenth-century medical thought and the emergence of new scientific ideas, institutions and practices.

THE INDIAN MEDICAL SERVICE

The Indian Medical Service (IMS) owed its institutional origins neither to metropolitan models nor to indigenous precedents, but arose out of the medical and military requirements of early colonial rule. Almost from the inception of the East India Company, surgeons had been assigned to ships travelling to India; from the second half of the seventeenth century they also served at the principal factories in India. With the territorial expansion of the Company in Bengal and southeastern India in the mid-eighteenth century the necessity arose for a permanent land-based establishment to meet the medical needs of Company servants and especially those of the army. The Bengal Medical Service, created in 1763, six years after the battle of Plassey, had an initial establishment of forty surgeons. Medical services on similar lines followed soon after in Madras and Bombay, and in 1785 the Court of Directors set the peacetime establishment for the three presidencies at 234 surgeons and assistant surgeons. By 1824, after decades of warfare and expansionism, the total had risen to 630. Thereafter, numbers fluctuated between about 650 and 820 for the rest of the colonial period.[1]

The bureaucratic structure of the service was strengthened by the establishment of provincial medical boards in the 1780s. Consisting of two or three senior surgeons in each presidency, these regulated appointments, maintained discipline and determined general policy. The system proved increasingly inefficient, however, and in 1857 each medical board was replaced by a provincial Director-General, later re-designated the Inspector-General or Surgeon-General.[2] In the 1860s provincial sanitary commissioners were added as part of the growing state commitment to public health but without executive powers of their own. This institutional structure was replicated at the centre: in the late nineteenth century the Government of India could draw upon the advice of its own Sanitary Commissioner as well as from 1895 a Director-General appointed to represent the Indian Medical Service. By the 1860s the IMS had matured into an archetypal colonial service, wedded to the military and administrative needs of the colonial state and staffed almost exclusively by Europeans specially recruited for the purpose. This was a model that paralleled (and so facilitated communication with) the structure of the Indian Civil Service (ICS) and the other scientific and technical services that emerged in the second half of the century. Despite conflict over policies and personalities that occasionally bordered on institutional civil war, the service structure promoted the development of a professional *esprit de corps* and a large measure of institutional conformity. The authority invested in the medical hierarchy could be effectively deployed to crush dissent and enforce compliance, as shown by the stifling of opposition to the anti-contagionist views held by

---

[1] D. G. Crawford, *A History of the Indian Medical Service, 1600–1913*, 1 (London, 1914), pp. 197–221.
[2] Roger Jeffery, *The Politics of Health in India* (Berkeley, CA, 1988), p. 61.

J. M. Cuningham, Sanitary Commissioner with the Government of India from 1866 to 1884.³

The military responsibilities of the IMS were evident from the outset. Members of the service held a double commission, as both medical and military officers, but it was repeatedly made clear in times of crisis that military needs took precedence over civilian duties. Under rules issued by Lord Cornwallis as Governor-General in 1788, surgeons were not permitted to enter civil employment until they had completed a stipulated period under the military: this was usually two years, spent with a regiment or at an army hospital. Officers were liable to be recalled from civil duties if the army required them, as many were during the First World War. Despite criticism of the IMS as anachronistic, and a resolution in the Indian legislature in 1918 calling for a separate Civil Medical Service, the Government of India insisted on the need to retain a single service that combined military and civilian responsibilities. The principal reason was to have a sufficient number of medical officers immediately available in case of war. But a further reason was to assist the recruitment and retention of Europeans in other state services, especially the ICS and the Indian Police. Given the paucity of European doctors outside the main cities, it was believed that the health (and peace of mind) of white civil servants and police officers, along with their wives and children, could be guaranteed only if a European medical officer were on call to assist them, even though, paradoxically, after 1858 the official duties of the IMS were confined to the Indian Army, while the medical needs of British soldiers were met by the Army Medical Department (later known as the Royal Army Medical Corps), which had an establishment in India of just under 300 in 1913. Although it was doubted by the 1920s that Europeans, even in cities like Calcutta and Allahabad, relied any longer on the services of white doctors alone, the Government of India still regarded racial identity as a valid reason for retaining a predominantly European IMS. A member of the Viceroy's Council reinforced this position in 1922 by arguing that failure to attract a 'reasonable' percentage of European recruits for the IMS would 'jeopardise the whole medical administration of the country to a dangerous degree'. He added that 'no one who has any knowledge of the very eminent benefits which have been conferred on this country by the Indian Medical Service in the past, could contemplate such a result without dismay'.⁴

---

³ J. C. Hume, 'Colonialism and Sanitary Medicine: The Development of Preventive Health Policy in the Punjab, 1860–1900', *MAS*, 20 (1986), pp. 703–24.

⁴ Sir William Vincent, Minute, 25 August 1922, Health and Education (Medical), 3–41, March 1923, NAI. For a subsequent public statement of this view, see *Indian Statutory Commission: Report*, 1 (London, 1930), p. 278.

Despite being 'essentially a military service', as India edged slowly towards a skeletal system of public health in the second half of the nineteenth century, it was officers from the IMS who were appointed to the most senior posts, including those of provincial and central sanitary commissioners, though few before the 1890s had any specific training in this field.[5] IMS officers also performed a wide variety of other administrative and technical duties, including managing prisons and mental asylums and filling professorial chairs in botany, chemistry, physiology and medicine. Nor were they confined to official duties. Until the 1830s Company surgeons could supplement their income with banking, commerce and landholding. Thereafter, except in posts that expressly forbade it, they drew an often substantial income from private practice, though by 1914 this was shrinking in the face of growing competition from independent (mostly Indian) practitioners.[6]

A strikingly large proportion of those who received posts in the service in the Company period and the early years of Crown rule came from Scottish medical schools and universities. This reflected the power of the patronage wielded by Scots, from the days of Henry Dundas as the first President of the Board of Control for seventeen years from 1784 onwards, in making appointments to military and civilian posts in India, but also the high standing of medical training in Edinburgh, Glasgow and Aberdeen. In the period 1839–60, just over half of all IMS recruits came from Scottish medical schools, and less than a third from English ones. But the balance steadily shifted so that, between 1897 and 1914, 43.9 per cent came from England, 27.0 per cent from Scotland, and 5.7 per cent from Ireland. As part of the reorganisation of the IMS, nomination was replaced by competitive examinations in 1855. In the second half of the century, entrants into the service were relatively well qualified compared with those entering either provincial practice in Britain or the British naval and military medical services.[7] Thereafter, however, concerns about promotion and status seem to have made the IMS less attractive to British medical graduates, a trend accentuated, so the Government of India believed, by the growth in Indian recruitment after 1905 and the changing political climate in India. The pressure for Indianisation and the interruption to European recruitment during the First World War significantly altered the racial composition of the IMS. Between 1913 and 1922 Indian recruits out-

[5] Mark Harrison, 'Towards a Sanitary Utopia? Professional Visions and Public Health in India, 1880–1914', *SAR*, 10 (1990), pp. 19–40.
[6] Jeffery, *Politics*, p. 64; C. P. Lukis, Minute, 21 September 1912, Home (Medical), 9–13, NAI.
[7] D. G. Crawford, *Roll of the Indian Medical Service, 1615–1930* (London, 1930), pp. 635–49; Mark Harrison, *Public Health in British India: Anglo-Indian Preventive Medicine, 1859–1914* (Cambridge, 1994), pp. 22–7.

numbered European by 122 to 111, prompting post-war government anxiety over the future of the service.[8] The IMS may have been slipping into 'terminal decline',[9] but, as will be seen in subsequent chapters, the promotion of tropical medicine and the opportunities for medical research were among the means by which the government and its senior medical officers sought to revive the prestige of the IMS and improve its attractiveness to European recruits. Indeed, the later history of the IMS demonstrates how tenaciously the colonial regime clung on to control of the Medical Service, even in the closing decades of British rule.

In general, the social and professional standing of medical officers in India appears not to have been very high, and certainly before 1857 they were often looked down upon by the European civilian and military elite. Apart from his 'empty pockets' and 'pallid countenance' (a third of all Company physicians died while on service in India between 1764 and 1838), the up-country civil surgeon was said to be much like 'his professional brother, the country general practitioner at home', but he was likely to find, on arriving at a new station, that 'his opinions and his practice' were treated, 'by the generality of his patients, with the greatest and most undisguised distrust'.[10] He was treated as a social inferior, and this 'low estimate of his position weakened his influence in the official sphere'.[11] And yet, however lowly the position of the up-country doctor, throughout the nineteenth century there were medical officers who attained positions of considerable political and professional influence or contributed in significant ways to the advancement of medicine and other scientific fields. Perhaps their very lack of status was an incentive for Company surgeons and their successors under the Crown to pursue an interest in botany, geology and other sciences in order to improve their standing in European society in India and to advance their career prospects.

INDIAN PRACTITIONERS OF WESTERN MEDICINE

Until late in the century, the Indian Medical Service essentially provided employment for Europeans. Indians had little opportunity to enter the IMS, still less to rise to senior positions in the service. Indians were essential, none the less, to the organisation and dissemination of Western medicine. Given the expense of training European doctors for service in India, their relatively high

---

[8] Vincent's Minute, 25 August 1922.
[9] Roger Jeffery, 'Recognising India's Doctors: The Establishment of Medical Dependency, 1918–39', *MAS*, 13 (1979), p. 310.   [10] 'Surgeons in India – Past and Present', *CR*, 23 (1854), p. 246.
[11] R. T. Lyons, *A Treatise on Relapsing or Famine Fever* (London, 1872), p. 253.

salaries, their wide range of administrative responsibilities, and the high levels of sickness and mortality among them, it was imperative from the standpoint of the Company to have in medicine, as in other branches of the colonial administration, Indian subordinates to perform routine duties. It was further considered unsafe to entrust the lives of Company servants to indigenous practitioners who had no training in Western medicine and the state had, therefore, to provide Indians with some appropriate form of medical education.

In Calcutta in 1824 and in Bombay two years later, training institutions were set up for this purpose, primarily designed to supply the army with sub-assistant surgeons, dressers and apothecaries. Calcutta's Native Medical Institution, which opened with twenty students, increasing to fifty in 1826, was to provide the Company with a regular supply of 'native doctors', taught through the vernaculars and through translations of English textbooks, but with parallel instruction in the indigenous medical systems. Classes in Ayurveda were also given at the Sanskrit College, using the works of Caraka and Susruta, while for Muslim students classes in Unani medicine were held in Urdu at the Calcutta Madrassa. A similar pattern of translation and vernacular instruction was followed at Bombay.

This apparent support for the Indian medical systems has led to some misrepresentation. It has been assumed that when the Calcutta Institution closed in 1835 it brought to an end an era of 'peaceful' cooperation and 'friendly' coexistence between the Western and Indian systems and signified the replacement of a benign Orientalist policy of patronising and learning from indigenous medicine by an intolerant Anglicist one, with disastrous consequences for the subsequent history of indigenous medicine.[12] But the differences between the two camps should not be exaggerated. As David Kopf pointed out, the main difference between the two was that 'Orientalists advocated syncretic schemes for self-help to one's own value system, whereas Anglicists were less sympathetic to traditional values and more convinced that no real change was possible without radical assimilation to the British style of life'.[13] With respect to medicine, official policy seems all along to have been directed to the practical goal of providing cheap but reliable medical aid for Company servants. The function of the Calcutta Native Medical Institution was never to promote indigenous medicine (which anyway formed a secondary part of the curriculum) as an equal or alternative to the Western system, but to 'train up a class of

---

[12] Brahmananda Gupta, 'Indigenous Medicine in Nineteenth- and Twentieth-Century Bengal', in Charles Leslie (ed.), *Asian Medical Systems: A Comparative Study* (Berkeley, CA, 1976), pp. 369–70; Poonam Bala, *Imperialism and Medicine in Bengal: A Socio-Historical Perspective* (Delhi, 1991), pp. 41–5.

[13] David Kopf, *British Orientalism and the Bengal Renaissance: The Dynamics of Indian Modernization, 1773–1835* (Berkeley, CA, 1969), p. 278.

native practitioners who would employ suitable native medicines with skill'. Offering instruction in Ayurveda and Unani medicine was also a ploy to attract recruits from the Vaidyas and other communities with a tradition of medical practice. Once recruited, it was assumed that they would come to recognise the superiority of Western medicine, even if they used cheap 'native remedies' instead of costly imported drugs in their professional work.[14]

In fact, the experiment of trying to combine elements of different medical traditions quickly proved unsatisfactory. Within a few years of its creation, Calcutta's Native Medical Institution was being criticised for the poor quality of its students and for failing to provide adequate instruction. The Bombay Institution closed in 1832, after only six years. In Bengal in 1833 the Governor-General, Lord Bentinck, appointed a committee for the purpose of 'improving the constitution and extending the benefits' of the Native Medical Institution and creating a system of management and education better suited to official needs. After a heated debate, the committee advised the abolition of the Institution, along with medical classes at the Madrassa and Sanskrit College, and the creation of a new college to teach Western medicine exclusively and with English as the sole medium of instruction. Bentinck approved these recommendations, and in 1835 the Native Medical Institution was replaced by the new Medical College.[15] Madras received its own medical college in the same year, but Bombay had to wait until Grant Medical College opened in 1845 for a replacement for its defunct Native Medical Institution. There is no doubting the symbolic, as well as practical, significance of the dissolution of the Native Medical Institution and the creation of the new Medical College in Calcutta. Even more than previously, Western medicine after 1835 was taken as the hallmark of a superior civilisation, a sign of the progressive intentions and moral legitimacy of colonial rule in India and the corresponding backwardness and barbarity of indigenous practice. **James Ranald Martin** compared the creation of the Medical College to Bentinck's other acclaimed act, the abolition of *sati* in 1829. Indeed, he rated the opening of the college as even more important 'in as much as the diffusion of European medical science, with its collateral branches, must prove one of the most direct and impressive modes of demonstrating to the natives, the superiority of European knowledge'.[16]

[14] Zhaleh Khaleeli, 'The White Man's Burden: The Rise and Fall of the Native Medical Institute, Calcutta', B.Sc. dissertation, University College, London, 1996, pp. 19–27; 'English Ideas, Indian Adaptation', *CR*, 30 (1858), p. 30; Mridula Ramanna, 'Indian Practitioners of Western Medicine: Grant Medical College, 1845–1885', *Radical Journal of Health*, 1 (1995), pp. 116–17.
[15] *Centenary of Medical College, Bengal, 1835–1934* (Calcutta, 1935), pp. 3–7.
[16] James Ranald Martin, *Notes on the Medical Topography of Calcutta* (Calcutta, 1837), p. 60.

Calcutta Medical College took over from its predecessor the task of producing Indian civil and military sub-assistant surgeons, vaccinators and dispensary attendants. Between 1835 and 1858 the college produced 456 'native doctors'. Not all entered Company service: some found employment under Indian princes and zamindars; others established private practices among the 'rich babus' of the city. Indians trained in Calcutta qualified as Licentiates in Medicine and Surgery; not until after the creation of Calcutta University in 1857 did they begin, in small numbers at first, to receive MD degrees.[17] Bombay's Grant Medical College, set up with financial assistance from the Parsi magnate Jamsetji Jeejeebhoy, served a similar function. Parsis formed the largest single contingent among early entrants to the College, forming more than 40 per cent of the intake between 1846 and 1866, but there were also substantial numbers of Christians and Hindus. It was assumed that Parsis and Christians were willing to take up medical training because they did not share caste Hindus' aversion to touching and dissecting human bodies, but if such an antipathy did exist it was not long in being surmounted. By the mid-1880s nearly a third of the students at Grant Medical College were Hindus, drawn mainly from the higher castes, especially Brahmins. Students came mostly from middle-class backgrounds, and included the sons of merchants, government servants and pensioners.[18] By the 1880s more than 300 doctors had qualified from the College. Through state service or private practice, Bombay's medical graduates contributed to the gradual dissemination of Western medicine in the province and in the neighbouring princely states, though as late as 1882 it was reckoned that only a tenth of the population of Bombay city received treatment from qualified practitioners of Western medicine.[19]

Across British India as a whole by the 1920s about 1,000 Indians were employed as assistant surgeons in the provincial (or 'subordinate') medical departments, backed by a further 4,000 sub-assistant surgeons. By contrast, although the IMS was opened to Indians with the advent of competitive examinations in 1855, it attracted few successful candidates. One of the first was S. G. Chuckerbutty, who had been sent to London for medical training in 1845, passed second in the examinations of 1855, joined the IMS and subsequently held the chair of materia medica and clinical medicine at Calcutta Medical College.[20] But, because of the strength of European opposition to Indianisation, only fifty-five Indians had joined the service by the outbreak of

---

[17] James Harrison, 'The Origin and Progress of the Bengal Medical College', *Indian Annals of Medical Science*, 5 (1858), p. 51; Anil Kumar, *Medicine and the Raj: British Medical Policy in India, 1835–1911* (Delhi, 1998), pp. 25–34. [18] Ramanna, 'Practitioners', pp. 118–20. [19] Ibid., p. 132.
[20] P. C. Sen Gupta, 'Soorjo Coomar Goodeve Chuckerbutty: The First Indian Contributor to Modern Medical Science', *MH*, 14 (1970), pp. 183–91.

the First World War, most of them since 1905.[21] Several Indian recruits resigned after relatively short careers, feeling themselves unwelcome or undervalued by European members of the service. Baman Das Basu resigned from the IMS in 1907, after only sixteen years' service, ostensibly because of ill health but in reality because he had 'too keen a sense of personal and national self-respect to relish being in harness ... with military imperialists'.[22]

For most of the century, the paucity of European doctors outside the IMS strengthened the service's hold over the medical profession in India. Equally, the fact that the IMS at first held the lion's share of private practice among European residents (as well as wealthier and more Westernised Indian families) reduced the prospects for the emergence of an independent medical profession. Perhaps because of high-caste Hindu taboos, or from the continuing strength of indigenous medicine in many parts of the country, but also because of the limited career prospects and financial rewards Western medicine offered, an Indian medical profession developed only gradually, more slowly than in the more lucrative and prestigious fields of law and government service. In the 1881 census there were said to be 'hardly any' qualified private practitioners in the Madras Presidency outside Madras city; by 1900 there were about 200, divided between Calcutta, Bombay and Madras, and of these about a third were Indian.[23] In 1908 the Secretary of State for India, Lord Morley, deplored the slow growth of an independent medical profession, mainly because of the financial burden this placed on the state to maintain the IMS. He saw the answer as increasing Indian recruitment into the IMS and removing a number of senior posts from the IMS as a bait to encourage independent doctors.[24] By the First World War, as enrolment in Indian medical schools and colleges began to soar,[25] a growing number of Indian men and women were entering the medical profession, finding employment with municipalities and local boards, commercial and state enterprises (such as railways, factories and tea estates), as well as in private practice. But this was still mainly in the cities, leaving the small towns and districts to folk medicine, to the practitioners of a revitalised indigenous medicine and to the few, widely scattered, public health institutions.

ENCOUNTERS WITH INDIAN MEDICINE

One explanation for the slow emergence of an independent Western medical profession was the continuing availability and perceived efficacy of indigenous

---

[21] Crawford, *History*, I, pp. 504–5; Jeffery, *Politics*, p. 64.
[22] Ramananda Chatterjee, 'Baman Das Basu', *MR* (December 1930), p. 668.
[23] Anil Seal, *The Emergence of Indian Nationalism: Competition and Collaboration in the Later Nineteenth Century* (Cambridge, 1968), pp. 120–1; *Thacker's Indian Directory, 1900* (Calcutta, 1900), pp. 376, 1202, 1327.
[24] Home (Medical), 37–55, December 1910, NAI.   [25] Jeffery, *Politics*, pp. 80–1.

medicine. To a degree unparalleled in other scientific fields (and only matched in a few areas of technology), the British were obliged to recognise the existence of older and culturally entrenched therapeutic beliefs and practices, and much historical debate has focused in recent years on the nature and consequences of the resulting interaction between Western and indigenous medicine and the extent to which the two constituted rival or complementary systems.

Early British attitudes to what was known of Ayurvedic, Unani and even folk medicine were often tolerant and even appreciative, though seldom without substantial qualification. This was partly because in their understanding of disease, as in their therapeutic devices, eighteenth- and early nineteenth-century Indian and European medicine shared much in common, not least an underpinning of humoral pathology. Although European physicians seemed loath to admit it, Western medicine remained a highly imperfect, empirical science, and for all the pride individual doctors showed in their own skills and nostrums, it was all too evident that when confronted with cholera or dysentery their medical chests lacked convincing cures. Even vaccination, a medical technology that largely worked, could not be scientifically explained in the absence of a more developed understanding of immunology. The British appreciated that there was much that they might usefully learn from indigenous medicine, particularly from its rich materia medica, accumulated over centuries of empirical trial and observation, but they had little time for its religious sanctions and cultural cosmologies. Given the cost and scarcity of imported drugs, such as the 'Peruvian bark' (cinchona) used to treat malaria, there were strong financial as well as therapeutic incentives to find local substitutes, and this was a major stimulus to the botanical investigations of Roxburgh and other early surgeon-botanists. To some extent British physicians were engaged in a typically Orientalist exercise, widening the bounds of Western knowledge by interrogating Oriental texts and 'native informants'. In 1814 the Court of Directors identified itself with this quest, noting (on the basis of reports received from India) that there existed in Sanskrit 'many tracts of merit... on the virtues of plants and drugs, and on the applications of them in medicine, the knowledge of which might prove desirable to the European practitioner'.[26]

But the tone was more often cautious, even sceptical. Sir William Jones doubted that there existed in any Asian language a single 'original treatise on medicine considered as a science', as opposed to a 'mere empirical history of

[26] Kopf, *British Orientalism*, p. 152.

diseases and remedies'.²⁷ The editor of the *Transactions of the Medical and Physical Society of Calcutta* observed in 1825 that the materia medica of the East had 'long contributed to the pharmacy of Europe' (a semantic distinction that itself implied the latter's superiority), and saw no reason to suppose that the Orient's storehouse had yet been exhausted. He averred, however, that the history of Indian medicine was likely to be 'of more interest than utility'. It was not to be expected that 'the imperfect science of the *Baids* and *Hakeems* of India' would offer 'any instructive lessons to their better educated brethren of Europe'. Still, 'to liberal and cultivated minds, the progress and condition of science in all ages, and in all climes, must be objects of interest', and the Society, accordingly, welcomed any light that might be thrown on 'the past or present existence of Oriental medicine, by information gathered from authentic sources, or derived from actual observation'.²⁸

Not until a generation after Jones, between about 1810 and 1830, did scholar-surgeons like Benjamin Heyne and Whitelaw Ainslie in Madras and H. H. Wilson in Bengal begin in earnest to study Ayurvedic texts. Their work was followed in 1838 by J. Forbes Royle's *Antiquity of Hindoo Medicine*, with its seminal claim for the anteriority of Hindu over Greek and Arabic medicine, and in 1845 by T. A. Wise's *Commentary on the Hindu System of Medicine*. With the partial exception of Royle, these authors displayed a critical attitude to their subject, especially what they saw (in Orientalist fashion) as the debased, present-day remnants of a system of medicine that, having flourished in ancient times, had long since fallen into deep decline. Ainslie, although impressed by the range and utility of India's materia medica and the skills of the physicians he encountered in Madras, had two grounds for reproaching Hindu medicine. The first was the way in which medicine had become mixed up with religion, so that Ayurveda was revered as a gift of the gods, 'a circumstance which has been an insurmountable obstacle to improvement' and a reason it was 'still sunk in such a state of empirical darkness'. Secondly, he regretted that the practice of dissection, referred to in ancient texts, had been abandoned and forgotten, thus leaving the Ayurvedic physician profoundly ignorant of the body's internal functions and disorders.²⁹ Wise, too, praised the 'advanced state' of learning in ancient India and (in a seemingly significant step towards medical pluralism) acknowledged Ayurveda as a scientific 'system' in its own right. But he contrasted ancient achievements with the complacency of modern physicians, who were 'satisfied with the knowledge and power . . .

---

[27] Girindranath Mukhopadhyaya, *History of Indian Medicine from the Earliest Ages to the Present Time*, I (Calcutta, 1923), pp. 4–5. [28] *Transactions of the Medical and Physical Society of Calcutta*, I (1825), pp. iii–iv.
[29] Whitelaw Ainslie, *Materia Indica*, II (London, 1826), pp. v, vii.

acquired at a very early period' and attached such 'blind reverence' to their ancient texts as to 'perniciously' retard any further advance.[30]

Nevertheless, at a time when other Indian sciences had fallen into growing disrepute among Europeans, there still appeared to be sufficient practical value to persist in the investigation of Indian medicine, whether through editing and translating texts or observing and conversing with *vaids* and *hakims*. The emphasis, almost invariably, was on 'useful' knowledge. Ainslie's *Materia Indica* in 1826 was one of the first attempts to establish 'a kind of combining link betwixt the materia medica of Europe and that of Asia'.[31] This was followed by articles in learned journals such as the *Transactions of the Medical and Physical Society of Calcutta*, by translations of texts, such as George Playfair's *Taleef Shereef* (the *Talifi-Sharifi*, an account of Indian drugs by the eighteenth-century *hakim* Sharif Khan) in 1833, and by the *Bengal Pharmacopoeia* compiled by **W. B. O'Shaughnessy** in 1844, which aimed to provide a guide to local substitutes for imported drugs and facilitate the search for new remedies in India, 'where a vast and rich field is open to the careful experimentalist'.[32] There was some willingness, too, to learn directly from Indian physicians. Twining, for instance, made a study of the malaria-related 'spleen diseases' of Bengal: he noted the drugs and therapies they employed and, following local practice, experimented with the use of iron needles to puncture swollen spleens. But, like many of his European contemporaries, he assumed that Western medicine already operated on a more securely scientific basis and that Indians acted only from 'practical knowledge' of the effects of the treatments they used, without much attendant understanding or reasoning.[33]

Twining was writing in the 1820s and 1830s when many European doctors still believed they could learn from Indian physicians, especially in the use of indigenous drugs and the treatment of diseases prevalent in India but uncommon in Europe. In fact, indigenous drugs and techniques were only one of several influences on Western therapeutic practice in early nineteenth-century India. In the treatment of malaria and cholera in particular, physicians deployed a range of 'heroic' measures, including copious bloodletting, violent purges and the extensive use of mercurials, that owed little, if anything, to indigenous practice but were adopted in the belief that the severity of disease in the tropics and

---

[30] T. A. Wise, *Commentary on the Hindu System of Medicine* (2nd edition, London, 1860), pp. i–v, xix.
[31] Ainslie, *Materia*, 1, p. x.
[32] W. B. O'Shaughnessy, *The Bengal Pharmacopoeia* (Calcutta, 1844), pp. iv–v; David Arnold, *Colonizing the Body: State Medicine and Epidemic Disease in Nineteenth-Century India* (Berkeley, CA, 1993), pp. 48–50.
[33] William Twining, *Clinical Illustrations of the More Important Diseases of Bengal*, 1 (Calcutta, 1832), pp. 332–4.

the rapidity with which fatal symptoms developed there demanded a degree of 'boldness' that would 'surprise the practitioner in more temperate countries'.[34] Thus, while decrying the use of 'dangerous' drugs such as arsenic and aconite by Indian physicians, European doctors were themselves engaged in acts of therapeutic violence that did more to swell the death-rate than effect lasting cures. By mid-century Western medicine had begun to shun such desperate remedies, but that did not necessarily promote a reconciliation with Indian medicine. Despite articles by Indian contributors appearing in the *Indian Medical Gazette* almost from its inception in 1866, urging the adoption of drugs used by *vaids* and *hakims* for conditions as diverse as malaria, dysentery and gonorrhoea,[35] the British appetite for cross-cultural therapeutic exchanges had greatly diminished. Western medicine felt increasingly secure in its superior knowledge, convinced that most that could usefully be gleaned from indigenous texts and informants had already been harvested. With the establishment of cinchona plantations in India, the quest for a local substitute for one of the most important drugs slackened, and European medicine was anyway shifting away from the endless accumulation of exotic materia medica to the pursuit of scientific pharmacology and the isolation of active chemical ingredients.

Even before this, little attempt was being made to differentiate between various forms of folk practice and the text-based traditions of Ayurvedic and Unani medicine. In 1837 J. R. Martin, whose medical topography helped shift the focus of Western medical attention away from encounters with Indian medicine to consideration of the influence of climate and landscape, denounced Indian physicians, without distinction, for their 'shameless impostures'.[36] A few years later Charles Morehead saw it as one of the foremost duties of the newly opened Grant Medical College to counter the 'demoralising effects of the irrational, superstitious and, too often criminal empiricism' of the followers of indigenous medicine. Their practice, he declared, was 'wholly unscientific' and consisted chiefly of 'charms, amulets, and incantations'.[37] There were partial exceptions to this trend of increasing outright condemnation, such as the drive in Punjab between the 1860s and 1880s to train *hakims* and their sons in the basic skills of Western medicine. But such schemes were contentious at the time and seen as no better than stop-gap measures,

---

[34] James Annesley, *Researches into the Causes, Nature, and Treatment of the Most Prevalent Diseases of India and of Warm Climates Generally*, 1 (London, 1828), p. xi.
[35] E.g., Meer Ushruff Alli, 'Treatment of Dysentery by Native Medicines', *IMG* (April 1868), p. 83; Chetan Shah, 'Gonorrhoea as Treated in Vernacular Medical Literature', *IMG* (September 1880), pp. 241–5. [36] Martin, *Topography*, p. 60.
[37] Hermann Haines, *Memorial of the Life and Work of Charles Morehead* (London, n.d.), pp. 15–16.

appropriate only until Western medicine had mustered the trained personnel and the financial and institutional resources to replace them.[38]

This is not, however, to deny that, despite the diatribes, a process of partial interaction and assimilation did occur. This can be shown, for example, through the career of Bhau Daji, a Maharashtrian who joined Grant Medical College in 1845, became an assistant surgeon, but left soon after to enter private practice in Bombay. In 1860 he opened a charitable dispensary with his brother, Narayan, also a practitioner of Western medicine, and with financial support from a Parsi philanthropist. Bhau Daji maintained close contacts with Grant Medical College and its Principal, Charles Morehead, writing papers on indigenous drugs and other medical topics for the College Medical Society. Despite Morehead's denunciation of indigenous medicine, the Society, founded in 1852, was an important forum for the exchange of medical information, especially about the properties of Indian drugs and their use in Western medical practice.[39] One of the first Indian members (and one-time vice-president) of the Bombay branch of the Royal Asiatic Society, Daji was also a keen Sanskrit scholar and used his knowledge of Ayurvedic texts to investigate indigenous drugs that might provide a cure for leprosy. In the course of this research, 'he discovered the ancient medicine, the Khasta, and tried it with remarkable success on his patients', but he did not reveal his secret before his death in 1874 and it was never 'scientifically' tested.[40] Daji straddled two worlds, combining, without apparent incongruity, the knowledge and techniques of Western medical science with what could usefully, and with pride, be retrieved from the Indian medical tradition. His interest in a cure for leprosy was indicative of a widely held belief among Indian practitioners that India's materia medica and Ayurvedic texts could yet yield remedies that had eluded the West. The adoption by Western doctors of chaulmoogra oil, possibly the basis for Daji's own experiments in the treatment of leprosy in the 1860s, and its widespread use into the early twentieth century, was taken as clear evidence that India still had a wealth of traditional knowledge to contribute to modern pharmacology.[41]

A further illustration of the eclectic exchanges and syncretic interface between Western and Indian medicine in the nineteenth century is provided by

---

[38] John C. Hume, 'Rival Traditions: Western Medicine and Yunan-i Tibb in the Punjab, 1849–1889', *BHM*, 51 (1977), pp. 214–31.   [39] Ramanna, 'Practitioners', pp. 126–8.

[40] T. G. Mainkar (ed.), *Writings and Speeches of Dr. Bhau Daji* (Bombay, 1974); obituary, *Journal of the National Indian Association* (August 1874), pp. 206–7.

[41] Jane M. Buckingham, 'Medicine and Confinement: Leprosy in Colonial South India', Ph.D. thesis, University of Sydney, 1996, pp. 107–81, 227–9; J. C. Ghosh, *A Monograph on Chaulmoogra Oil and Its Use in the Treatment of Leprosy as Explained in Ayurveda* (Madras, 1917).

quinine. Extracted from the bark of the cinchona tree, a native of Andean South America, quinine had no traditional place in Indian medicine and attempts by the British to introduce it as a prophylactic and treatment for malaria met with widespread resistance in India. It was held to have 'heating' properties and unpleasant side-effects (Indians were not alone in their aversion to the latter, one reason why many Europeans were also antipathetic to its use). Nevertheless, the efficacy of quinine in the treatment of malaria did gain unofficial recognition in indigenous medical practice. In the 1860s it was being used as a basic ingredient in mixtures prepared by Bengali *kavirajs* for the treatment of 'Burdwan fever', the epidemic malaria then sweeping central Bengal. It was also extensively used among *hakims* in Punjab.[42] By the 1920s large quantities of quinine, sold at subsidised rates through post offices to encourage its use by the rural poor, were being bought up to be incorporated and sold, at a profit, as an ingredient in indigenous febrifuges.[43] But, it might be added, cinchona was not altogether exceptional, despite its remote origins. When it came to the use and incorporation of foreign drugs, Indian pharmacology had a long tradition of eclecticism. Ayurveda had borrowed extensively from Unani medicine in the past, sarsaparilla root had been introduced from America to treat syphilis, and several of the main ingredients used in Ayurvedic medicine in the 1890s were not native to India but imported from the Middle East, East Africa, and Central and Southeast Asia.[44]

SMALLPOX: FROM VARIOLATION TO VACCINATION

The impact of British attitudes on indigenous medical ideas and practices can most clearly be seen in the case of smallpox. Here the main target of attack was not Ayurveda or Unani but the widely disseminated folk practice of variolation. This involved inoculation with live smallpox matter, as opposed to vaccination (using vaccine originally derived from cowpox), which was introduced into India by the British. Since it was one of the few forms of medical intervention in nineteenth-century India that (for all its defects) could confidently be said to work, vaccination both had great practical importance and served to demonstrate the growing sense of superiority of Western medicine over indigenous practice.

[42] Gopaul Chunder Roy, *The Causes, Symptoms and Treatment of Burdwan Fever or the Epidemic Fever of Lower Bengal* (London, 1876), p. 106; C. A. Gill, 'Malaria: A Scientific Romance', *Proceedings of the Lahore Philosophical Society*, 1 (1914), p. 28.
[43] C. A. Bentley, *Malaria and Agriculture in Bengal: How to Reduce Malaria in Bengal by Irrigation* (Calcutta, 1925), p. 92.
[44] *Report of the Central Indigenous Drugs Committee of India*, 1 (Calcutta, 1899), appendix 3, pp. 10–19.

There are several accounts of variolation in India by Western observers from the mid-eighteenth century onwards. Most of those written before 1800 were broadly favourable, whereas those produced subsequently, after the introduction of Edward Jenner's technique of vaccination, were largely condemnatory. One of the most detailed and frequently cited accounts was that given by J. Z. Holwell to the College of Physicians in London in 1767. It was based on observations made during his residence in Bengal and was intended to reassure the British of the safety and effectiveness of inoculation at a time when it was still a novelty in Britain, being recently introduced from the Middle East and regarded with some suspicion. Holwell described the variolators as 'a particular tribe of Bramins', who travelled early each year from north India, arriving in the villages of Bengal shortly before the smallpox season began in January or February. They normally inoculated on the patient's outer arm: having rubbed the chosen spot for several minutes with a piece of cloth, the inoculator pricked the skin with a small metal instrument and wrapped onto the arm a wad of cotton impregnated with 'variolous matter' taken from pustules produced by the previous year's inoculations. According to Holwell, 'they never inoculate with fresh matter, nor with matter from the disease caught in the natural way, however distinct and mild the species'. The use of old and therefore attenuated viral matter ensured that the disease was not reproduced in its full severity and yet had sufficient potency to produce a prophylactic effect. Holwell stressed the care with which variolators excluded pregnant women, unprotected adults and others who might be vulnerable to smallpox, and he commended the 'cool regimen' of cold water douches and 'cooling' items of food and drink used to counter the fever induced by inoculation. The smallpox produced by variolation was, Holwell assured his audience, of a mild nature: the number of pustules was small and there was little risk either of the disease spreading epidemically or its causing the death of those inoculated. Despite the 'multitudes' inoculated every year in Bengal, the practice 'adds no malignity to the disease taken in the natural way, nor spreads the infection, as is commonly imagined in Europe'.[45]

Although Holwell and his contemporaries were unable to give a precise indication of how widely variolation was practised in India at the time, it appears from later evidence that in Bengal and neighbouring areas at least 60 per cent of the population was protected in this way. Variolation was a craft practised by many different castes, not just the Brahmins Holwell identified, and in parts of northern India as well as in Bengal. The British themselves took up the

---

[45] J. Z. Holwell, *An Account of the Manner of Inoculating for the Small Pox in the East Indies* (London, 1767).

practice in Calcutta and elsewhere in the late eighteenth century in the hope of protecting themselves against one of the most feared and fatal of all Indian diseases. Nor did variolation disappear rapidly after the British introduced vaccination to India in 1802. As late as 1850 there were reported to be 42 inoculators or *tikadars* ('mark-makers') in Calcutta alone and early attempts to outlaw the practice or replace it with vaccination met with scant success.[46] However, variolation did come under increasing attack from British physicians. In 1831 Dr W. Cameron blamed *tikadars* in and around Calcutta for recurring epidemics of smallpox and for spreading 'falsehoods and ridiculous stories' about vaccination. The Smallpox Commission of 1850, in comparing variolation to *sati* and infanticide, declared that the time had come to suppress 'this murderous trade'.[47]

More resolute action was taken in the vaccination acts of the 1870s and 1880s to outlaw variolation and make vaccination compulsory in such designated areas as municipalities and cantonments. But, even in the 1860s, there were individuals, such as T. E. Charles, Bengal's Superintendent-General of Vaccination, who argued that the evils of variolation had been exaggerated; it was, he believed, rarely the cause of epidemics or of mortality among the inoculated, and until vaccination was more widely available it provided the populace with a valuable means of protection. He even proposed that variolators be licensed and brought under state control. This was unacceptable to the government, but a number of former *tikadars* were recruited in Bengal in the 1870s to carry out vaccination for the state, though the suspicion lingered that they secretly practised variolation instead.[48] Ultimately, by co-opting *tikadars* and outlawing their trade, by making vaccination more widely available and overcoming initial resistance to it, the British were able by the close of the nineteenth century to effectively suppress a once widespread practice.

Variolation was not the only source of resistance to vaccination. Smallpox was widely identified with the Hindu goddess Sitala, whose awesome presence was manifested through the disease's fever and eruptions. A benign outcome to 'possession' by the goddess was sought through songs, prayers, devotional offerings and cooling potions.[49] Although worship of the goddess was not integral to the inoculators' craft, her assistance was often tactfully invoked by variolators as they began their work. By contrast, the practice of vaccination

[46] Arnold, *Body*, pp. 129–33.
[47] W. Cameron, 'On Vaccination in Bengal', *Transactions of the Medical and Physical Society of Calcutta*, v (1831), p. 387; *Report of the Small-Pox Commissioners* (Calcutta, 1850), p. 54.
[48] *Bengal Vaccination Report, 1869*, pp. 1–11; Arnold, *Body*, pp. 146–7.
[49] Ralph W. Nicholas, 'The Goddess Sitala and Epidemic Smallpox in Bengal', *JAS*, 41 (1981), pp. 21–44.

was seen as ungodly and offensively polluting in its crude transmission of body fluids from one individual to another. Belief that Sitala was being defiled or assaulted contributed to Indian distrust of vaccination and nurtured rumours about the selfish and cynical reasons that motivated British vaccinators and their assistants to want to put their 'mark' (*tika*) on Indians. Vaccination, like later measures to curb cholera and plague, thus formed an important site of cultural resistance to colonial medical intervention.[50]

Nevertheless, from the British perspective smallpox was too dangerous a disease to be ignored, and the lives of European soldiers and civilians would continue to be threatened unless effective measures of control were introduced. There was, besides, an initial (if misguided) expectation that Indians would welcome vaccination as a product of the cow Hindus held sacred and as singular evidence of the magnanimity of British rule. Once Jenner's discovery in 1798 of the prophylactic effect of cowpox inoculation was conveyed to India in 1802, Europeans rapidly switched from inoculation to vaccination. The initial rejection of vaccination by many Indians was seen by colonial physicians as proof of their depravity and perverse ingratitude, their blind adherence to 'ancient usage' and the 'degrading' effects of the Hindu religion.[51] However, as with many another imported technology, the British were gradually forced to modify the practice of vaccination to suit local cultural and environmental conditions. Rather than expecting vaccination to be taken up voluntarily, the state had to recognise that it would have to bear the main costs of its dissemination itself and to create a suitable vaccinating agency. The nature of this agency varied from one part of British India to another. In Bengal much reliance was placed on public dispensaries and on winning over erstwhile variolators to vaccination. In Bombay, by contrast, attempts were made to carry vaccination directly to the people. In 1827 the province was divided into four vaccination circles, each with its own team of itinerant Indian vaccinators, acting under European supervision. The vaccinators visited each village in their area at least once a year and gave free vaccination to as many children as possible. This scheme was later taken up in Punjab, Madras and elsewhere. It also came to be recognised that during certain seasons of the year, when heat and humidity turned vaccination marks into 'foul sloughing sores', vaccination was impractical in India. It took decades of experimentation, too, before the quality of the vaccine, much of it initially imported from Britain, became reliable in Indian conditions and before sufficient quantities of vaccine

---

[50] Arnold, *Body*, pp. 143–4; Frédérique Apffel Marglin, 'Smallpox in Two Systems of Knowledge', in F. A. Marglin and S. A. Marglin (eds.), *Dominating Knowledge: Development, Culture, and Resistance* (Oxford, 1990), pp. 112–18.   [51] Arnold, *Body*, pp. 134–7.

could be produced locally. The abandonment of arm-to-arm vaccination, in favour of vaccination directly from calves or using lymph preserved with lanolin or glycerine, made vaccination not only more effective, but less coercive and painful and, without the direct transmission of body fluids, culturally more acceptable. By the latter half of the nineteenth century vaccination was acquiring influential support from Indian practitioners of Western medicine and from prominent members of the Indian middle classes in places such as Bombay.[52]

Only once these technical modifications had been made and India was producing sufficient quantities of its own vaccine could the scale of operations begin to match the scale of the disease itself. From a mere 350,000 vaccinations in British India in 1850, the number rose to 4.5 million in 1877, by which time there were more than 2,500 full-time vaccinators. By the 1890s the annual number of vaccinations had risen to nearly 8 million and by early in the twentieth century to 9 million.[53] In many provinces between a third and a half of all children were being vaccinated; in Bombay the figure was 80 per cent. That, however, still left a fifth of all children unprotected and, despite legislation, initiated in Bombay in 1877 and taken up by the Government of India in 1880, to make vaccination compulsory in designated areas, this left a reservoir of susceptible individuals large enough to generate fresh outbreaks of smallpox. None the less, in one of the most striking measures of medical intervention and one of the most significant modifications of India's disease environment in colonial times, vaccination was beginning by 1914 to make appreciable inroads into smallpox mortality. Epidemics were becoming less frequent and less intense, and though smallpox had not been 'conquered', and remained unvanquished until the 1970s, it was a far less important cause of Indian mortality by the First World War than it had been even fifty years earlier.

MALARIA AND MEDICAL TOPOGRAPHY

British attitudes to variolation demonstrated the capacity of Western medicine to veer from a degree of borrowing to a direct confrontation with Indian medical practices and beliefs. Much of the colonial medical science of the nineteenth century, however, directed attention away from cultural engagements to encounters with the Indian environment. Although in the land of

---

[52] Mridula Ramanna, 'Indian Response to Western Medicine: Vaccination in the City of Bombay in the Nineteenth Century', in A. J. Qaisar and S. P. Verma (eds.), *Art and Culture: Endeavours in Interpretation* (Delhi, 1996), pp. 67–78.

[53] S. P. James, *Small-pox and Vaccination in British India* (Calcutta, 1909), pp. 30, 76–8.

Sitala its cultural context might be strikingly different, as a disease smallpox in India was not seen to be very different from smallpox in Europe. By contrast, many of the diseases prevalent in India were believed to draw their distinctive character, and their exceptional potency, from the peculiarities of the Indian climate and landscape, though sometimes, as in the case of cholera, the effect of environmental idiosyncrasies – such as the monsoon – were seen to be aided and abetted by such peculiar rites and customs as Hindu pilgrimages and bathing festivals. Through topographical surveys and accounts of the physiological and pathological effects of 'warm climates', India was defined as an exotic space, a dangerous and unfamiliar place, largely unsuited, even in the 'temperate' hill-stations, for permanent white settlement.[54] Although necessarily obeying the same 'universal' laws as in Europe, disease in India appeared to function in unfamiliar or extreme ways. The theory and practice of medicine had, accordingly, to admit to local 'modifications' and these at times opened up a significant gulf between colonial and metropolitan medical science.

The idea that the physical environment exercised a potent influence on human health and disease was an ancient one. It could be traced in European thought at least as far back as Hippocrates in the sixth century BC, and in India a well-established theme of environmental determinism ran through Ayurvedic thought and therapeutic practice.[55] Although there is little evidence to show that nineteenth-century British medical writers were much aware of this indigenous tradition, they sometimes cited Indian opinion in support of their own views as to the healthiness or insalubrity of a given location. Of more immediate influence was the role of neo-Hippocratic and environmentalist ideas in eighteenth- and early nineteenth-century Europe, especially following publication of Montesquieu's *De l'esprit des lois* in 1748, and the growth of medical topography as a way of relating the observed incidence of diseases, especially epidemics, to season, vegetation, rivers, marshes and other physical or environmental factors.[56]

In their accounts of the diseases of seamen, visiting fleets and coastal stations, early medical accounts of India, even as late as the 1790s, reflected the external, essentially maritime, nature of British contact with India.[57] After

---

[54] Arthur S. Thomson, 'Could the Natives of a Temperate Climate Colonise and Increase in a Tropical Country and Vice Versa?', *Transactions of the Medical and Physical Society of Bombay*, IV (1843), pp. 112–38; James Murray, 'Practical Observations on the Nature and Effects of the Hill-Climates of India', ibid., VII (1844), pp. 3–30.

[55] Francis Zimmermann, *The Jungle and the Aroma of Meats: An Ecological Theme in Hindu Medicine* (Berkeley, CA, 1987).

[56] James C. Riley, *The Eighteenth-Century Campaign to Avoid Disease* (Basingstoke, 1987).

[57] John Peter Wade, *A Paper on the Prevention and Treatment of the Disorders of Seamen and Soldiers in Bengal* (London, 1793).

about 1800, however, as British rule became more established, physicians began to compile more systematic accounts of the diseases of India and to relate them to the experience of the West Indies and other 'warm' or 'tropical' climates. Of particular significance in its dual emphasis upon climate as a source of disease and the vulnerability of Europeans in the tropics was James Johnson's *The Influence of Tropical Climates, More Especially the Climate of India, on European Constitutions*, first published in 1813. This long remained a highly regarded text, a revised edition appearing as late as 1856.[58] Physicians with greater Indian experience than Johnson were by this time applying ideas of medical topography derived initially from Europe and the West Indies to their unfolding knowledge of India's diseases and the manner in which these affected Indians as well as Europeans. From the late 1820s the *Transactions* of the Calcutta Medical and Physical Society (and, later, its Bombay namesake) carried a steady stream of articles on medical topography, most compiled, like the geological surveys of the period, during official tours of duty or in the course of military marches. Along with more comprehensive accounts of the diseases of Madras and Bengal, these constituted a preliminary mapping of India's disease environment, especially for the broad category of 'fevers' (most of which were probably attributable to malaria).

In the 1820s James Annesley of Madras argued that medicine in India could be put on a 'rational footing' only if due attention were paid to climate, season and the geographical distribution of disease. His own 'topographical and statistical reports' on the Madras Presidency were a preliminary attempt to put this idea into practice.[59] In 1835 James Ranald Martin in Calcutta won government approval for a series of local medico-topographical reports, which extended and systematised the work of earlier surveys by men such as Francis Buchanan. Martin's own *Notes on the Medical Topography of Calcutta* in 1837 helped define the genre, and was closely followed by such texts as John M'Cosh's *Topography of Assam* (1837) and Robert Rankine's *Notes on the Medical Topography of the District of Sarun* (1839). Works of this kind continued into the latter half of the century when they were superseded by the district gazetteers. The broad approach they adopted, encompassing geology, climate, vegetation, diet, disease and indigenous medicine, reflected the breadth of the surgeons' scientific interests but also made medicine central to the investigation and representation of the Indian environment. As befitted a medical agency so closely tied to military needs, data about disease in the army (along with the

---

[58] On Johnson's influence, see Mark Harrison, 'Tropical Medicine in Nineteenth-Century India', *BJHS*, 25 (1992), pp. 299–318.
[59] James Annesley, *Sketches of the Most Prevalent Diseases of India* (London, 1825), pp. 251–5.

jails) provided one of the few sources of information available to medical statisticians like Annesley. Equally, one of the principal uses of medical topography was to seek ways to curb high levels of mortality and morbidity among European troops by relocating barracks or directing troop marches away from notoriously unhealthy swamps and jungles. Medical topography was instrumental, too, in the reconnaissance of upland India, from the Nilgiris to the Himalayas, in the search for sanatoria for sick and disabled troops and for European civilians anxious to escape the heat and diseases of the plains.[60] Although much attention was directed to diseases that constituted a threat to European health, medical topography also brought to light information about disorders that rarely affected Europeans and whose connection with environmental factors remained unclear. A notable example of this was the discussion of goitre in the Himalayas, including work by the surgeon-geologist John M'Clelland incorporated into his *Medical Topography* of Bengal and the North-Western Provinces in 1859.[61]

The medical investigation of climate and topography helped fashion a new, more censorious, attitude to India. This was nowhere more so than in relation to Bengal, which only decades earlier had been seen as a land of remarkable abundance, an Indian Eden. The famine of 1770, the subsequent epidemics of cholera and malaria, and the decline of the indigenous textile industry all contributed to increasingly negative representations of Bengal, but medical topography, too, cast its dismal pall over the province. Invoking the miasmatic understanding of disease then current, F. P. Strong in the 1830s admitted he could not explain exactly how 'malaria' was formed but had no doubt it was 'produced most abundantly in all those parts of Bengal which are not cleared of jangal [jungle], drained, and kept clean'. Even around Calcutta, there were all the 'essentials necessary for the formation of malaria' – 'lakes, marshes, gardens crowded with trees. . . . stagnant water, filthy pools, and low grass jangals of every kind'. Aided by the 'natural heat and moisture of the climate', these generated a constant supply of the malaria 'poison'. When, in addition, 'unnatural or meteoric changes of the climate' occurred, or there were 'unnatural inundations of sea, or river water', 'disease and death scourge[d] the land'.[62] In his account of Dacca in 1840, James Taylor identified similar environmental influences that made the countryside 'abound with malaria'. The

---

[60] *Report on the Medical Topography and Statistics of the Ceded Districts* (Madras, 1844); *Report on the Medical Topography and Statistics of Neilgherry Hills* (Madras, 1844). On hill-stations generally, see Dane Kennedy, *The Magic Mountains: Hill Stations and the British Raj* (Berkeley, CA, 1996).

[61] M. Miles, 'Goitre, Cretinism and Iodine in South Asia: Historical Perspectives on a Continuing Scourge', *MH*, 44 (1998), pp. 50–6.

[62] F. P. Strong, *Extracts from the Topography and Vital Statistics of Calcutta* (Calcutta, 1837), pp. 9–10.

sickly season started as the annual floods receded in September, and from then until late November malaria was at its height, leaving some localities so badly affected that the inhabitants existed 'in a state of perpetual fever'. The effects on the inhabitants of 'this gloomy region', with its 'noxious atmosphere', were evident in their 'sallow, cadaverous looks, tumid bodies and shrunk emaciated limbs'.[63] Twining, too, situated Bengalis in a morbid landscape, contrasting their physique with that of the inhabitants of northern India, whom he considered 'a stronger and more hardy race'. He attributed this difference in part to the Bengalis' rice diet, but also to 'the climate which is damp and unfavourable to health, the rainy season being prolonged, and the atmosphere contaminated by exhalations from swamps and districts intersected by rivers and creeks'.[64]

These negative medical representations of Bengal and Bengalis were further intensified by the spread from the mid-1850s of 'Burdwan fever', a prolonged epidemic (most likely to have been malaria) which caused heavy mortality and provoked renewed pessimism about rural Bengal.[65] The annual reports of the provincial sanitary commissioners and the decennial census reports from 1871–2 onwards gave detailed and depressing evidence of the destruction wrought by malaria. It was not just the large numbers of dead that attracted comment, but the many more who were incapacitated by the disease. Malaria was represented as an 'emasculating' disease that threatened reproduction, rendered individuals weak and sickly, and so accentuated the division, already entrenched in colonial ideology and practice, between the 'manly' and 'martial' races of the north and northwest and the 'effeminate' Bengalis.

It is barely necessary to sample this medico-ethnographical discourse to understand the authority medicine brought to colonial representations of race and why Bengalis themselves internalised such representations in seeking to explain their own physical and political weakness. In 1889 Bengal's Sanitary Commissioner calculated that 'malarial fever' was responsible for three-quarters of all mortality in the province, amounting to almost 1 million deaths a year. He urged the local authorities to do more to combat this menace so that in time those 'wretched beings of sallow and ghastly countenance, looking twice their real age, with attenuated frames, shrunken limbs, [and] muscles thin and powerless' would be replaced by 'men well-knit, with their muscles developed, and their vital organs sound – altogether powerful, vigorous, healthy and

---

[63] James Taylor, *A Sketch of the Topography and Statistics of Dacca* (Calcutta, 1840), pp. 322, 329–31.
[64] Twining, *Illustrations*, II (2nd edition, Calcutta, 1835), p. 418.
[65] J. Elliot, *Report on the Epidemic Remittent and Intermittent Fever Occurring in Parts of Burdwan and Nuddea Divisions* (Calcutta, 1863).

happy'.[66] Four years later the Census Commissioner, C. J. O'Donnell, painted an almost apocalyptic picture of the desolation caused by malaria. From Calcutta north virtually to Darjeeling there existed 'a large area of decaying or nearly stationary population'. In one district 'the spectacle of whole villages depopulated by a brooding mortality' was 'almost universal'; in another 'ruined houses and abandoned sites were everywhere visible'. The effects of fever were so widespread that 'the sickly physique' had become 'the ordinary physique'.[67] Year by year, added L. S. S. O'Malley in 1911, malaria was 'silently and relentlessly at work'. Not only did it 'diminish the population by death'; it reduced 'the vitality of the survivors', sapping their 'vigour and fecundity'.[68]

In the face of such insistently negative representations, and all the statistics and medical testimony that supported it, it is not surprising that middle-class Bengalis adopted a gloomy view of their environment and its human consequences. Echoing the environmental determinism of the English historian H. T. Buckle, the civil servant and historian R. C. Dutt remarked of Bengal in 1873: 'All those physical causes which enfeeble and enervate, and make man incapable of having mastery over Nature, are found to exist and work in this country to an alarming extent.'[69] In 1876 a Bengali doctor, Gopaul Chunder Roy, published his own mournful account of 'Burdwan fever', and his funereal images were repeated again and again in Bengali writing over the next fifty years.[70] In the wake of the census reports of 1891 and 1901, the view became widespread among middle-class Bengali Hindus that theirs was a 'dying race', decimated by malaria, while the Muslims of eastern Bengal, where malaria was less prevalent, continued to multiply. This powerful association of disease with race and religion was influentially articulated by an Indian IMS officer, U. N. Mukherji in 1909, and became one of the issues behind mounting Hindu–Muslim tension in Bengal, though in the ensuing debate factors other than malaria (such as diet and marriage customs) were also adduced to explain why Muslims were a more 'vigorous' race and were gaining a demographic and political advantage over Hindus.[71]

Of course, malaria was more than a matter of racial representation. Although the cause of malaria remained elusive until Ross's research in the

---

[66] W. H. Gregg, 'Malarial Fever in Bengal', *CR*, 88 (1889), p. 384.
[67] *Census of India, 1891, III: The Lower Provinces of Bengal* (Calcutta, 1893), pp. 2, 64, 88, 92.
[68] *Census of India, 1911, V: Bengal, Bihar and Orissa, and Sikkim* (Calcutta, 1913), p. 519.
[69] 'Arcydae', 'The Past and Future of Bengal', *Bengal Magazine* (January 1873), p. 251.
[70] Roy, *Fever*; Upendranath Brahmachari, *Gleanings from My Researches*, 1 (Calcutta, 1941), pp. 473–83.
[71] David Arnold, '"An Ancient Race Outworn": Malaria and Race in Colonial India, 1860–1930', in Waltraud Ernst and Bernard Harris (eds.), *Race, Society and Medicine, 1700–1960* (London, 1999), pp. 123–43.

mid-1890s revealed the link between the malaria parasite, mosquitoes and human infection, during the course of the century a vast amount of material about the diagnosis, treatment, prevention and distribution of the disease was assiduously assembled and used, at times, to practical effect. For all its erroneous assumptions, even miasmatic theory had a value in prompting the relocation of human habitations away from malarial sites or in raising questions about the healthiness of colonial irrigation and railway-building projects. As early as the 1850s T. E. Dempster was able to demonstrate a connection between malaria and the spread of canal irrigation. He also devised the 'spleen index', a reliable guide to the intensity and distribution of the disease.[72] But the history of malaria and its close association with medical topography were no less important in informing British and Indian attitudes towards environment, health and race, and in establishing the authority of medicine in representations of the self and the other.

CHOLERA: ENVIRONMENT, ARMY AND PUBLIC HEALTH

Like malaria, cholera was one of the most formidable diseases of nineteenth-century India. It caused in the region of 15 million deaths between 1817 and 1865, and yet, until the end of the century, its origins remained obscure. When a devastating epidemic of the disease broke out in the Jessore district of Bengal in 1817, spread throughout South Asia and, a decade later, invaded the West, many European observers believed that they were witnessing the birth of a new disease. However, there had been outbreaks of this, or a very similar disease, in India over the previous forty years, and epidemics recorded centuries earlier by the Portuguese and Dutch may also have been of cholera. But the fact that cholera was specifically identified with India made it a matter of particular concern to doctors there. The scale of the mortality involved, the speed with which the disease spread and the apparent randomness with which it chose its victims made this one of the most feared, but least explicable, diseases of the nineteenth century.

There was little consensus among medical writers in India as to how cholera was caused or transmitted. Early hypotheses linked cholera to 'atmospheric vicissitudes', such as heavy downpours of rain or a sudden fall in temperature, or, like malaria, to miasmatic factors, such as the 'poisonous emissions' and 'pestiferous exhalations' from rotting vegetation, crowded habitations and

---

[72] T. E. Dempster, *The Prevalence of Organic Disease of the Spleen as a Test for Detecting Malarious Localities in Hot Climates* (Calcutta, 1868).

human 'filth'. James Jameson was one of the first to point to atmospheric disturbances in his account of the 1817 outbreak, but he saw these as 'accessory' factors rather than a specific cause.[73] As the 'father of Indian cholera literature', his work had an enduring influence on subsequent writers in India and this possibly explains why so many of them clung to climatic or environmental explanations long after these had been jettisoned in Europe. But, as with malaria, theories of cholera aetiology and transmission revealed a sustained determination by colonial India's medical establishment to situate disease in the idiosyncrasies of Indian climate and topography (and culture), even when the absurdity of this was pointed out by doctors in Europe, who were confronted with epidemic cholera in their own towns and cities. Cholera exemplified the localist contention, spelt out by Morehead in 1838 and repeated with approval by Sir Joseph Fayrer more than forty years later, that 'disease in India is not disease in England'.[74] Distancing India from Europe in this way was a measure of the almost hypnotic strength of Western physicians' engagement with what they understood to be the distinctive nature of the Indian environment. It also established a defiant camaraderie among doctors, who, as professional exiles from Europe, were anxious to assert their own authority by arguing that only those who knew India from years of hard-earned personal experience could possibly pronounce upon the nature of its diseases or devise measures appropriate to their containment.

One explanation advanced in the second half of the century was that cholera epidemics in India were related to the passage of the monsoon, a view particularly advocated by the Government of India's Statistical Officer, James L. Bryden, in 1869. Sceptical that cholera could be transmitted by contaminated water, as proposed by John Snow in London, Bryden argued that human agency mattered less than meteorology. He used extensive statistical data to support his claim that the prevailing winds determined the direction and extent of cholera epidemics in India: 'The highways by which cholera travels are, in this country, aerial highways, and not routes of human communication.' The pattern of cholera outbreaks could thus be confidently attributed to 'the special physical geography of Hindustan' and the 'perfection of the regularity with which the seasons come forward year after year'.[75]

Alternative – or supplementary – explanations were not entirely lacking. It came to be recognised that humans played some part in the spread of the

---

[73] James Jameson, *Report on the Epidemick Cholera Morbus* (Calcutta, 1820), pp. 27, 87.

[74] *Transactions of the Medical and Physical Society of Bombay*, I (1838), pp. iv–x; Joseph Fayrer, *On the Climate and Fevers of India* (London, 1882), p. 154.

[75] James L. Bryden, *Epidemic Cholera in the Bengal Presidency* (Calcutta, 1869), pp. 91–2, 243.

disease, even if it could not be established exactly what. Focusing on the peculiarities of Indian religious culture as much as on its physical environment, Hindu festival sites and pilgrimage routes were identified as one of the primary means by which cholera epidemics were, if not generated, then at least widely disseminated. Epidemic outbreaks coinciding with the twelve-yearly Kumbh Mela, or bathing festival, at Hardwar in 1867 and again in 1879 were taken to exemplify this lethal connection. Puri in Orissa, Pandarpur in Maharashtra and several pilgrimage sites in south India were similarly implicated.[76] In another important link to social and environmental factors, famine was also identified as having a critical role in the spread of the disease. W. R. Cornish, the Madras Sanitary Commissioner, demonstrated with respect to the famines of 1866 and 1876–8 that the 'abnormal conditions' to which the malnourished, migrant poor of the countryside were exposed, the dearth of clean water and proper food, and their congregation in insanitary towns and relief camps made them particularly vulnerable to cholera as well as being agents in its wider dispersal.[77]

But Bryden's meteorological arguments and the belief in the local causes of cholera outbreaks enjoyed the overwhelming support of the Government of India's Sanitary Commissioner, J. M. Cuningham. He argued, as in the case of the 1879 Hardwar outbreak, that pilgrims fell prey to cholera because their fatigue and privations made them 'specially prone', not because they carried contagion from place to place. He believed that cholera was probably caused by an 'aerial miasma' or generated by some obscure condition of the air or soil, or possibly both together.[78] Although supporting the need for local sanitary measures, Cuningham denied the value of wider measures to control fairs and pilgrimages, a view that the Government of India, wary in the wake of the 1857 uprising of interfering in Indian religious practices, was predisposed to accept. It also suited the government that Cuningham opposed the sanitary cordons and quarantine measures that other European states demanded but that the British opposed as an unwarranted constraint on the freedom of trade. As long as he remained in office, Cuningham (who retired in 1884) used his authority to smother criticism and suppress contagionist views.[79]

In one of the first investigations of its kind in India, T. R. Lewis of the Army Medical Department and D. D. Cunningham of the IMS were deputed in 1868

---

[76] *Report on the Cholera Epidemic of 1867 in Northern India* (Calcutta, 1868); *Report Regarding the Control of Pilgrimages in the Madras Presidency* (Madras, 1868).
[77] W. R. Cornish, *Cholera in Southern India* (Madras, 1871), pp. 1–4; *Review of the Madras Famine, 1876–1878* (Madras, 1881). [78] J. M. Cuningham, *Report on the Cholera Epidemic of 1879 in Northern India* (n.d.).
[79] Mark Harrison, 'A Question of Locality: The Identity of Cholera in British India, 1860–1890', in David Arnold (ed.), *Warm Climates and Western Medicine: The Emergence of Tropical Medicine, 1500–1900* (Amsterdam, 1996), pp. 135–46.

to examine the aetiology and spread of cholera in India. They used the latest laboratory techniques, including microscopy, but failed to break away from the dominant belief in local causes. They concluded that human agency alone could not explain the peculiar distribution and periodicity of the disease. The cholera 'poison' developed neither in water nor in the human body, but in the soil, a medium that was in turn influenced by climate and by seasonal variations in rainfall and humidity. Cholera, they concluded, had 'as good a claim as malarial diseases to a telluric origin'.[80] Another medical officer, H. W. Bellew, who conducted his own statistical investigation into cholera in the 1880s, similarly rejected contagionist theory, seeing instead 'a fixed relation between cholera and special climatic conditions'.[81] Even when in 1884 the German bacteriologist Robert Koch identified the comma bacillus as the cause of cholera and located it, under the very noses of the Indian medical establishment, in a water tank (reservoir) in Calcutta, it was several years before his explanation gained general acceptance in India, so entrenched was the belief in environmental causes and influences.[82]

Although cholera caused enormous mortality and suffering among the population at large, it was seen to have a particularly severe impact upon European troops in India, and this military consideration was one of the main incentives for its intensive investigation and the prolonged quest for effective countermeasures. Between 1818 and 1854 more than 8,500 British soldiers were reported as having died of cholera, and recurrent outbreaks were a major factor in raising the average annual mortality among white troops during the period to a staggering 69 per 1,000. The military's affliction swelled to crisis proportions in 1861 with a fresh eruption of cholera in northern India: nearly 2,000 British soldiers, their wives and children were attacked, and about two-thirds of them died. The shock of this mortality and fears for the future health and security of the British Army in India (its importance enhanced by the increased number of British troops stationed there after the Rebellion of 1857) forced the authorities to take seriously the question of how to control cholera.[83] Even before that, high mortality among British troops in India had caused growing concern. In the wake of sanitary reform in Britain and following vigorous lobbying by Florence Nightingale, a Royal Commission was

---

[80] T. R. Lewis and D. D. Cunningham, *Cholera in Relation to Certain Physical Phenomena* (Calcutta, 1878), p. 115; Jeremy D. Isaacs, 'D. D. Cunningham and the Aetiology of Cholera in British India', *MH*, 42 (1998), pp. 283–8.   [81] H. W. Bellew, *The History of Cholera from 1862 to 1881* (London, 1885), p. v.
[82] Isaacs, 'Cunningham', pp. 291–8.
[83] *Report of the Commissioners Appointed to Inquire into the Cholera Epidemic of 1861 in Northern India* (Calcutta, 1863).

appointed in 1859 to enquire into the sanitary state of the army in India. The appointment of the Commission was an important precedent, the first of several similar bodies appointed over the next eighty years to investigate and report on such matters as plague research, irrigation, industrial development and agriculture. The demise of the Company exposed India to direct British scientific scrutiny and gave metropolitan experts such as William Farr, Britain's leading medical statistician and a member of the 1859 Commission, a greater influence over Indian policy. Unlike most of its successors, however, the sanitary Commission did not visit the subcontinent, though it did take evidence from old India hands like Martin.

But not all the reforming influences came from London, nor were they solely directed towards the preservation of European health. Ideas of state medicine and sanitary reform had been circulating in India for decades. Some doctors were beginning to regard improved public health as one of the fundamental duties of a modern, civilised state; others, more pragmatically and with an eye to attracting official support, stressed the benefits to the colonial regime of presiding over a healthier, and hence more stable and productive, population. In a work ostensibly devoted to European health, Martin called in 1856 for a system of state medicine that would promote 'increased security of life and property' through 'sustained and organised investigation, by competent persons, of all those agencies and circumstances, moral and physical which deteriorate, through local or general influences, the public health'.[84]

When the Commission reported in 1863, it duly gave priority to sanitary measures designed to improve white soldiers' health, but recognised that the danger from epidemics was shared by Europeans and Indians alike. Britain, it averred, had a moral obligation to help its Indian subjects by giving them the benefits of Western civilisation, and among these sanitation was one of the most prized. As Nightingale saw it, 'The next great work ... is sanitary reform in India', and since in India 'government is everything' that required greater and more determined action on the part of the colonial state.[85] In 1864, after further pressure from London, a sanitary commission was appointed for each Indian province, replaced in 1866 by provincial sanitary commissioners. In theory, this marked a vital stage in the emergence of state responsibility for public health in India, and suggested the importance, as in Britain, of cholera in impelling sanitary reform. But the immediate results were most evident in

---

[84] James Ranald Martin, *The Influence of Tropical Climates on European Constitutions* (London, 1856), pp. 103–4. [85] Edmund A. Parkes, *A Manual of Practical Hygiene* (London, 1864), p. 563.

the army, where, due to better hygiene and sanitation rather than improved medical treatment, mortality among white soldiers fell steadily over the next thirty years and by 1899 had sunk to under 15 per 1,000.[86] Among the Indian population, however, cholera mortality remained devastatingly high: in 1900, a severe famine year, over 800,000 deaths from cholera were recorded. Despite Nightingale and Martin, the colonial state was unwilling to shoulder the financial and political responsibility for wide-ranging public health measures. The sanitary commissions were 'purely investigative and advisory bodies'; the commissioners who replaced them, too, were merely advisers with no executive powers of their own.[87] As the form and content of the sanitary commissioners' annual reports reveal, the health of the army remained the greatest priority, and commissioners lacked the financial resources and technical staff that would have enabled them to develop effective public health programmes. Instead, responsibility for public health was largely relegated to the municipal councils and local boards that were coming into existence in India in the 1880s. These bodies, too, lacked the funds, the trained medical personnel and sanitary staff, and often the political will to carry out expensive and technically demanding sanitation schemes.

By the 1890s the colonial government faced mounting criticism for its failure to provide an effective health system for India. Although significant advances had been made in the health of British (and, latterly, Indian) soldiers, the country as a whole was entering a period of severe epidemic mortality, 'a woeful crescendo of death' caused by successive bouts of cholera, plague and malaria, which pushed the death rate in India above 40, and briefly even 50, per 1,000. Successful public health policies, Ira Klein has argued, 'would have shielded the populace from epidemics'. Instead an 'imperfect understanding of illnesses, financial strains, rural poverty, and weak municipal laws on sanitation all resulted in a tatterdemalion public health movement'.[88] Medicine and public health in India had fallen far behind Europe and North America and had failed to cope with the health problems produced by rural poverty and famine, by urban growth and spreading slums, and by the disease-generating effects of railways and canals. Public health and sanitation, Klein concludes, were 'the ugly ducklings of a civil service which rewarded political and military competence ... far more highly'.[89]

[86] Philip D. Curtin, *Death by Migration: Europe's Encounter with the Tropical World in the Nineteenth Century* (Cambridge, 1989), pp. 104–29.

[87] Radhika Ramasubban, 'Imperial Health in British India, 1857–1900', in Roy MacLeod and Milton Lewis (eds.), *Disease, Medicine and Empire: Perspectives on Western Medicine and the Experience of European Expansion* (London, 1988), p. 43.

[88] Ira Klein, 'Death in India, 1871–1921', *JAS*, 32 (1973), pp. 639–41.   [89] Ibid., p. 656.

WOMEN, MISSIONS AND MEDICINE

The state was not the sole patron of Western medicine in India, nor its only agency, though for much of the nineteenth century it remained the dominant presence. Christian missions, too, saw great potential value in medical work, and their involvement in medicine, specifically in women's medicine, is a reminder of the way in which medicine could also serve and inform wider agendas of religion and gender. Missionary involvement in the medical field began soon after the lifting of the Company's ban on Christian missionaries in 1813. From the 1830s, the London Missionary Society in south India began to take up medical work as part of its evangelising activity, believing that medicine could 'open a wide and effectual door into the hearts and minds of the natives', if only because medical aid was 'one of the very few forms of help which the Hindu is at liberty to receive'.[90] But medicine brought male missionaries disappointingly few converts, and, initially through the involvement of their wives and daughters, they began to see the possibility of using women nurses and doctors as a more effective agency. Medicine was one of the few scientific and technical fields open to women in nineteenth-century India, as in the West, partly because women were seen to have a natural proclivity for caring for and nursing the sick, but also because, in the Indian context, they were thought to be able to penetrate the 'dark' recesses of the hitherto uncolonised zenana (women's quarters), gain an influence over the women of the household, and through them reach their husbands and sons.[91]

In the 1850s the Zenana Bible and Medical Mission began sending women missionaries and later 'lady doctors' to the women's quarters of Hindu and Muslim households. The Delhi Female Medical Mission, a branch of the Society for the Propagation of the Gospel, began work in 1867; two years later, in 1869, the American Methodist Episcopal Mission sent Clara Swain to Bareilly, the first fully qualified woman medical missionary to be employed in India, and opened the first hospital exclusively for women and children in India. Fanny Butler, one of the first cohort of students at the London School of Medicine for Women, arrived in India in 1880 on behalf of the newly formed Church of England Zenana Missionary Society.[92] The opening up of medical education for women in the West (combined with the relative paucity of posts available to them there) facilitated the recruitment of medical women

---

[90] R. Lovett, *The History of the London Missionary Society, 1795–1895*, II (London, 1899), p. 223.
[91] Janaki Nair, 'Uncovering the Zenana: Visions of Indian Womanhood in Englishwomen's Writings, 1813–1940', *Journal of Women's History*, 2 (1990), pp. 13, 22.
[92] Margaret I. Balfour and Ruth Young, *The Work of Medical Women in India* (London, 1929), pp. 16–18.

for India in the second half of the century. Because it was widely believed that social taboos prevented male doctors (especially white male doctors) from treating Indian women, particularly those from the higher castes and classes, India offered women doctors a promising field of employment. Although medical education was slowly opened up to women in India (beginning with Madras Medical College in 1875), early recruits were drawn disproportionately from the Europeans, Eurasians, Parsis and Indian Christians; relatively few high-caste Hindu and Muslim women qualified as doctors before 1914.[93] Private schemes, like the one launched in 1883 by an American businessman, George A. Kittredge, and his Parsi associates, brought a small number of independent women doctors to India, including Edith Pechey, the chief physician at the new Cama hospital for women in Bombay. But the main initiative remained with the missionary societies. The task of their women doctors was to effect a 'double cure' – the healing of spiritual as well as physical 'disease'.[94] In addition to domiciliary visits to purdah households, they worked in the growing number of dispensaries and hospitals for women and children. By the 1890s, there were at least fifty women missionary doctors from various denominations, representing two-thirds of all female physicians in India at the time. By the late 1920s their number had risen to around 150 and mission hospitals for women constituted more than half of all such hospitals in India. The first medical school for women, opened at Ludhiana in Punjab in 1894 as the Women's Christian Medical College, was also designed to train Indian women as medical missionaries.

But the state did not remain entirely absent from the women's medical scene. In 1885, at the prompting of Queen Victoria, the Vicereine, Lady Dufferin, launched the National Association for Supplying Medical Aid to the Women of India. The Association sought to recruit qualified white women doctors, nurses and midwives to work in India, to give medical relief through hospitals, wards and dispensaries exclusively for women and children, and to educate and train Indian women for the medical profession. The setting up of the Dufferin Fund has long been seen as a turning-point in the history of Western medicine in India: while increasing women's access to medical facilities and bringing them medical relief, it also widened its own hitherto constricted social parameters and contributed to the marked surge in the number of patients attending hospitals and dispensaries in the closing decades of the nineteenth century. The Dufferin Fund also stimulated other initiatives in women's health and

[93] Kumar, *Medicine*, pp. 39, 43, 54–62.
[94] Rosemary Fitzgerald, 'A "Peculiar and Exceptional Measure": The Call for Women Medical Missionaries for India in the Later Nineteenth Century', in Robert A. Bickers and Rosemary Seton (eds.), *Missionary Encounters: Sources and Issues* (London, 1996), p. 185.

medical education, including the establishment of an all-India women's college, named after another vicereine, Lady Hardinge, at Delhi in 1916. Although criticised at the time by nationalists for favouring white women doctors, the Fund provided unique opportunities for Indian as well as European women to enter, and advance their careers in, the medical profession. But it has also been seen in a far more critical light. Not only is it said to have had a 'limited' impact on women's health (because of its small number of medical staff and its widely scattered facilities); it was also 'a lucid example of British paternalism in India'. By stereotyping Indian women as being intrinsically ignorant, backward, oppressed by their own menfolk and dependent on European initiative for their moral and physical well-being, the Fund constituted (so Maneesha Lal has argued) the single most important factor in the 'institutionalisation of gender' in the history of colonial medicine in India.[95]

A further drawback to the Dufferin Fund was that, despite its independent, philanthropic status, it remained perilously dependent on state favour. Founded by a vicereine, blessed by the Queen-Empress, run largely by officials and their wives, the Fund was inevitably tied institutionally and politically to the colonial power. But the state was reluctant to take on direct responsibility for women's health, and this left the Fund in a curious limbo between government aid, private philanthropy and missionary medicine. By 1910 the Dufferin Fund had 13 provincial branches, 140 local committees, and 160 hospitals, wards and clinics under its charge, but it was also in deep financial and administrative difficulties and could not survive without closer association with the state. Having earlier rejected calls for a women-only service, the Government of India finally relented and in 1914 authorised the setting up of the Women's Medical Service, financed through a state subsidy to the Dufferin Fund of Rs 370,000 a year. The new service was considerably smaller than the IMS, beginning with a cadre of 25 in 1914, rising to 44 (and a reserve of eight) in 1927, compared with the nearly 800-strong IMS. Moreover, since its duties were confined to the care of women and children, it was entrusted with a far more limited range of medical and administrative responsibilities than the IMS.

The 'colonising' of women as medical subjects was not confined to the efforts of Christian missions and the Dufferin Fund. Recent scholarship has revealed two other areas in which the medical profession laid singular claim to authority over women's bodies. The first of these, as an adjunct to military

---

[95] Meredith Borthwick, *The Changing Role of Women in Bengal, 1849–1905* (Princeton, NJ, 1984), pp. 324–6; Dagmar Engels, 'The Politics of Childbirth: British and Bengali Women in Contest, 1890–1930', in Peter Robb (ed.), *Society and Ideology: Essays on South Asian History* (Delhi, 1993), p. 233; Maneesha Lal, 'The Politics of Gender and Medicine in Colonial India: The Countess of Dufferin's Fund, 1885–1888', *BHM*, 68 (1994), p. 30.

medicine, concerned the creation of 'lock hospitals', for the confinement and treatment of prostitutes suspected of harbouring venereal diseases, at cantonments throughout India. Begun around 1805, these hospitals were officially suppressed in 1833, but reinstated in 1868 under the Contagious Diseases Act, before being technically repealed again in the 1880s but covertly continuing under new cantonment acts. It is striking that, at a time when few medical facilities existed for women, or even for the Indian public at large, the perceived medical and disciplinary needs of the army – to protect British troops from an infection supposedly spread by Indian prostitutes and to reduce high levels of hospitalisation from venereal disease among the white soldiery – could give rise to such wide-ranging medical powers in the cause of medically disciplining Indian women.[96] It has been argued that the impact of the Contagious Diseases legislation in India went still further, introducing, beyond the perimeters of the cantonments themselves, 'a new form of bodily regulation', in which the 'moral division between respectable and unrespectable women in India began to be detached from a sacred social hierarchy and became, instead, expressed through Western medical metaphors of health and disease'.[97]

A second area of medical intervention centred on moves to reform Indian midwives (*dais*) or, alternatively, replace them with Western-trained midwives. In the second half of the nineteenth century, *dais* came to be portrayed as 'wizen hags', whose 'barbaric' and 'primitive' practices brought suffering and death to women in childbirth and drove infant mortality to horrendous levels. The issue of midwifery seemed, like the activities and pronouncements of the Dufferin Fund (and of Western medicine in nineteenth-century India in general), to pit a progressive, humane West against a cruel and backward East; the *dai* thus became 'a symbol of [Indian] superstition and dogged resistance to change'.[98] And yet, at the same time, it is possible to see the attack on the *dais* and attempts to reform Indian obstetrics as more than an aggressive, one-sided act of cultural imperialism – not least because it had many Indian supporters. In Calcutta, as in Punjab, and doubtless in many other of the cities and provinces of India, the physical management of childbirth was beginning to change as a result of the increasing influence of Western 'medicalisation', and

---

[96] Philippa Levine, 'Venereal Disease, Prostitution and the Politics of Empire: The Case of British India', *Journal of the History of Sexuality*, 4 (1994), pp. 579–602; Douglas M. Peers, 'Soldiers, Surgeons and the Campaigns to Combat Sexually Transmitted Diseases in Colonial India, 1805–1860', *MH*, 42 (1998), pp. 137–60.

[97] Judy Whitehead, 'Bodies Clean and Unclean: Prostitution, Sanitary Legislation, and Respectable Femininity in Colonial North India', *Gender and History*, 7 (1995), p. 41.

[98] Geraldine Forbes, 'Managing Midwifery in India', in Dagmar Engels and Shula Marks (eds.), *Contesting Colonial Hegemony: State and Society in Africa and India* (London, 1994), p. 171.

so, too, were wider perceptions of women and reproduction.[99] As in the discussion of race and malaria alluded to earlier, or in the controversy over the Age of Consent Bill of 1891,[100] the ideas and authority of Western medical science, mediated through various religious and social reform movements, had by the late nineteenth century become an integral and influential part of Indian debates about gender, class, community and nation. Western medicine in India was no longer merely a colonial project.

[99] Supriya Guha, 'A History of the Medicalisation of Childbirth in Bengal in the Late Nineteenth and Early Twentieth Centuries', Ph.D. thesis, University of Calcutta, 1996; Anshu Malhotra, *'Pativratas* and *Kupattis*: Gender, Caste and Identity in Punjab, 1870s-1920s', Ph.D. thesis, SOAS, London, 1998, ch. 6.

[100] Himani Bannerji, 'Age of Consent and Hegemonic Social Reform', in Clare Midgley (ed.), *Gender and Imperialism* (Manchester, 1998), pp. 21–44.

CHAPTER 4

TECHNOLOGIES OF THE STEAM AGE

It is a well-established convention to see the nineteenth century as an age of innovative steam technologies, developed first in Europe, then diffused to other regions of the globe. In an era characterised in terms of 'a massive transfer of technology from the West to Africa and Asia', and with colonialism as a convenient conduit, India has often been taken to exemplify the momentous scale and impact of this process.[1] But although the transfer of technology argument duly highlights the importance of exogenous innovation and the role of technology as a 'tool of empire',[2] it can easily become an excessively one-dimensional idea, stressing the dynamism of the West but ignoring the context in which new technologies were employed. Technologies are seldom discrete bodies of knowledge, transferable wholesale, without emendation, from one society to another. Technological transfers are more likely to take the form of a 'dialogue' rather than a simple process of diffusion or imposition,[3] and this was especially the case in India, which had a wide range of existing technologies and a physical and social environment far removed from that of Europe. Equally, it needs to be recognised that under colonialism the dissemination of new technologies was constrained and conditioned by the partisan nature of political and economic control. Colonial rule interrupted the 'inventive exchanges' India had formerly had with its Asian and Indian Ocean neighbours and created instead a near-monopoly of technological dialogue with and through the West, and, primarily, with Britain itself. State power was used to promote technologies that served the regime's military, economic or ideological needs while restricting Indian access to technologies that might harm metropolitan interests. Denial was often as important as diffusion in shaping the technology of colonial India.

A further reservation about the impact of technological transfers from the West arises from the survival and adaptation of pre-steam technologies. In many instances, seemingly competing technologies – the old and new in agriculture, textiles, transport and labour – nestled side by side, enjoying a relation-

[1] Daniel R. Headrick, *The Tentacles of Progress: Technology Transfer in the Age of Imperialism, 1850–1940* (New York, 1988), p. 6.
[2] Daniel R. Headrick, *The Tools of Empire: Technology and European Imperialism in the Nineteenth Century* (New York, 1981).
[3] Arnold Pacey, *Technology in World Civilization: A Thousand-Year History* (Oxford, 1990).

ship that was more synergistic than antagonistic. Despite the tendency of some scholars to see the technology of the age as having a simple, unequivocal message, equating steam power with progress and engines with empire, in India the new technology was the bearer of mixed messages. The meaning attributed to steam technology by Western minds was not necessarily that communicated to Indian observers and participants, and among Europeans and Indians alike there were doubts about the desirability of modern technology and its suitability for India. The social and political cost of technological change remained a continuing source of debate and raised awkward questions about the meaning and value of India's modernity.

TEXTILES, TECHNOLOGY AND TRADITION

The history of India's cotton textile industry has often been taken as the most critical illustration of how steam supplanted craft production. Nationalist writers attributed the precipitous decline of traditional modes of textile manufacture in the early decades of the nineteenth century to competition from British cotton mills, which first cut off India's export markets and then flooded India itself with cheap goods against which handloom production could not compete. The result was to drive impoverished weavers (along with other displaced artisans) into the already crowded ranks of agricultural labourers. Through poverty and pressure on the land, this process of 'de-industrialisation' was seen as having contributed to the devastating famines of the nineteenth century. Even the rise of India's own mill industry after 1850 failed to provide an adequate alternative source of employment. By 1914 cotton was one of India's leading industries in terms of the capital invested and the number of mills and workers employed, but it was an industry plagued by technological backwardness and constrained by foreign competition.[4]

Recent scholarship has, however, amended this interpretation in a number of ways. There is no doubting that a major process of technological change occurred, or that its form owed much to outside influences. The majority of textile mills in India followed British designs and housed British machinery. More than 90 per cent of the textile machinery for cotton and jute mills imported into India between the 1850s and 1930s came from Britain, particularly from Lancashire, and between 1905 and 1939 alone this trade was worth

[4] D. R. Gadgil, *The Industrial Evolution of India in Recent Times, 1860–1939* (5th edition, Delhi, 1971); A. R. Desai, *Social Background of Indian Nationalism* (4th edition, Bombay, 1966), chs. 4–6. For recent reassessments, see Colin Simmons '"De-industrialisation": Industrialisation and the Indian Economy, c. 1850–1947', *MAS*, 19 (1985), pp. 593–622, and Rajnarayan Chandavarkar, 'Industrialization in India before 1947: Conventional Approaches and Alternative Perspectives', ibid., pp. 623–68.

Rs 780 million. India accordingly gained relatively few benefits for its own industrial technology, and, despite the enormous growth of the industrial sector, a machine- and tool-making industry had barely begun to emerge in India before 1939.[5] None the less, despite the dominance of imported machinery and the presence of European overseers and technicians, India's textile mills also to some extent reflected local conditions and pre-existing work practices. For instance, in a situation where there was a wide social and cultural gulf between mill-hands and managers, the jobber played a critical role as an intermediary in the recruitment and management of factory labour. Pre-factory skills and labour divisions also found some place in the composition and organisation of mill-workers, as in the case of the Julahas, Muslim weavers from north India, recruited to work in the cotton mills of Bombay.[6]

The most substantial way in which the conventional picture of 'de-industrialisation' has been revised has been by investigating the long-term adaptation and survival of handloom weaving in India. For centuries before the advent of colonial rule, India had been renowned for its textiles and particularly for the quality and colourful variety of its cotton goods. Until the late eighteenth century it was 'probably the world's greatest producer of cotton textiles'.[7] Domestic taste and external demand stimulated production, fostered specialisation, and gave some encouragement to the spread of new technologies, such as the spinning-wheel or *charka* (possibly introduced from Iran as late as the fourteenth century). This labour-saving device helped speed up the process of spinning, thus making larger quantities of yarn available for weaving; it also produced yarn of a more even quality than could be spun by hand.[8] Textiles in turn stimulated ancillary technologies, among them the making of wooden printing blocks and the production of dyes and mordants, such as alum extracted from copper-mine waste. With the 'craze' for Indian calicoes and other dyed, painted and embroidered textiles in the West from the second half of the seventeenth century, and with a growing export trade in European hands, cloth from Bengal, Gujarat and the Coromandel coast became a prized staple of the European East India companies.

---

[5] Robert Kirk and Colin Simmons, 'Lancashire and the Equipping of the Indian Cotton Mills: A Study of Textile Machinery Supply, 1854–1939', in Kenneth Ballhatchet and David Taylor (eds.), *Changing South Asia: Economy and Society* (London, 1984), pp. 169–81.

[6] Richard Newman, *Workers and Unions in Bombay, 1918–1929: A Study of Organisation in the Cotton Mills* (Canberra, 1981), pp. 35–57; Rajnarayan Chandavarkar, *The Origins of Industrial Capitalism in India: Business Strategies and the Working Classes in Bombay, 1900–1940* (Cambridge, 1994), chs. 6–7.

[7] K. N. Chaudhuri, 'The Structure of Indian Textile Industry in the Seventeenth and Eighteenth Centuries', in Tirthankar Roy (ed.), *Cloth and Commerce: Textiles in Colonial India* (New Delhi, 1996), p. 33.

[8] Irfan Habib, 'Technological Changes and Society, 13th and 14th Centuries', in Debiprasad Chattopadhyaya (ed.), *Studies in the History of Science in India*, II (New Delhi, 1982), pp. 818–22.

However, the success of the Indian textile industry lay not in machinery but in low-cost labour, in the abundance of raw cotton, and in the manual dexterity and 'empirical and hereditary knowledge' of spinners, weavers and dyers. The limits of this technology were exposed well before British industrialisation precipitated its decline: subject to the influences of climate and season, faced with ever-increasing demand but with such limited technical resources at their command, weavers and dyers found it difficult to maintain the volume, quality and variety of cloth sought by European merchants. The paucity of machinery in textile production (compared with China as well as Europe) and the reliance on hand technology are reasons for arguing that India was far from being on the brink of an industrial revolution of its own in the seventeenth or eighteenth centuries.[9]

In eighteenth-century Bengal, which, for all its commercial ascendancy, may have been 'technologically primitive' even compared with other Indian textile-producing areas, looms were made of wood and followed a very basic design. They appear to have changed little in recent centuries; even the use of a foot-lever or treadle seems to have been absent.[10] It required twenty separate stages to transform raw cotton into finished cloth; each depended on its own group of caste specialists using only elementary tools. For instance, yarn was produced, mainly by women, using a hand-cranked cotton-gin, a carding comb made from a fishbone, a bamboo teasing bow with a silk or catgut string, and a spindle or spinning-wheel. Dressing the cloth, mostly men's work, also called for several labour-intensive processes – washing, bleaching, repairing and beetling. Dyeing, painting and embroidering, which added greatly to the export value of the cloth, involved several further skilled but time-consuming processes. The quality of the cotton goods produced, accordingly, was 'inversely related to the technical sophistication employed': the finer the cloth the greater the input of labour required.[11] Heavily dependent upon, and indebted to, Indian middlemen, the weavers had no capital of their own to invest in improving production, nor was there a technologically minded class of superior artisans and entrepreneurs to experiment with, invent or improve machinery as in eighteenth-century Britain. The weavers' low income gave their cloth a competitive advantage but left them acutely vulnerable to subsistence crises.[12]

While making some effort from the 1760s to promote silk-winding

[9] Chaudhuri, 'Industry', pp. 75–6.
[10] Tirthankar Roy, 'Introduction', in Roy (ed.), *Cloth and Commerce*, p. 23; Habib, 'Changes', p. 844a; Hameeda Hossain, *The Company Weavers of Bengal: The East India Company and the Organization of Textile Production in Bengal, 1750–1813* (Delhi, 1988), plates 8 and 9.   [11] Hossain, *Weavers*, pp. 21, 38.
[12] Chaudhuri, 'Industry', pp. 69–72.

machines (filatures) in Bengal to improve the quality of country silk and compete with Italian production in Europe,[13] the East India Company showed no equivalent interest in trying to introduce the innovations in cotton-spinning and weaving that were revolutionising textile production in Britain. As a result, armed with the advantages of economy of scale and in the absence of any tariff barriers to protect India's handloom industry, the mass production of cotton cloth from British mills rapidly displaced Indian cloth from its former export markets by the 1820s and proceeded to capture a large share of India's domestic market as well. Exports of cotton piece-goods, worth Rs 165 lakhs in 1816–17, sank to a mere 8 lakhs in 1830–31, while imports in the same period rose from Rs 3 lakhs to 60. Innovations in transport – steamships, the opening of the Suez Canal in 1869, and the growth of India's railway system – intensified the impact of imported textiles over the following four or five decades, destroyed the old textile industry in Bengal, and crippled that in many other parts of India.[14] This inaugurated a new era of technological dependence on the West – first in cloth and subsequently in textile-making machinery. It is important to recognise, though, that this was not an unqualified tale of 'de-industrialisation', for the handloom industry did not die out entirely, but survived through a process of adaptation and specialisation that left it much reduced in size but technologically more sophisticated than it had previously been.

In most areas of India during the nineteenth and early twentieth centuries there was a long-term decline in the number both of weavers and of looms. Village weavers tended to disappear, especially once the railways brought a flood of cheap imported goods; those who survived tended to congregate in a small number of specialised urban centres. Some south Indian weavers switched to coarser varieties of cloth, to try to undercut foreign competition, and, despite market fluctuations and famines, survived well into the 1870s. Whereas the market for plain cloth fell to British and later Indian mills, handloom weavers in places like Salem and Sholapur maintained a hold through specialised types of cloth, especially those needed for ceremonial purposes or for dhotis, saris and bodices. As well as assisting the inroads of imported cloth, railways could also help handloom weavers to market their own goods more widely; the migration of Indian labourers overseas created new export markets for handloom lungis. The availability of yarn from foreign and, after the 1850s, Indian mills gave handloom weavers access to larger and more reliable

---

[13] Sabyasachi Bhattacharyya, 'Cultural and Social Constraints on Technological Innovation and Development: Some Case Studies', in Surajit Sinha (ed.), *Science, Technology and Culture: A Study of the Cultural Traditions and Institutions of India and Ceylon in Relation to Science and Technology* (New Delhi, 1970), pp. 44–72.   [14] *Report of the Fact-Finding Committee (Handloom and Mills)* (Delhi, 1942), p. 5.

quantities of their basic material, while technological innovations such as the fly-shuttle and the jacquard loom (a century after its appearance in Europe) added speed and consistency of quality and design to handloom production. After 1920 electric power brought further benefits and helped sustain the 'dynamics of continuity' in the handloom industry. This, then, was no mere passive survival: indeed, as late as the 1930s a third of India's total textile production came from handlooms.[15]

Though historians have substantially revised the history of India's 'deindustrialisation', it is as well to note that the image of the handloom did not always keep pace with changing reality. Textile production was one of the aspects of Indian technology that most attracted European interest in the late eighteenth and early nineteenth centuries, at a time when the possibility still existed of learning from Indian techniques. The investigation of India's 'arts' (signifying its crafts and technical skills) figured prominently alongside its sciences in Sir William Jones's inaugural address to the Asiatic Society in 1784. Three years later, Anton Hove set out on his investigative tour of the cotton districts of Gujarat. In 1791 Sir Joseph Banks remarked that nothing was 'more interesting to the progress of arts in Europe than communication from well inform'd men of the minute practice of Indian workmen'.[16] But contemporary accounts of Indian crafts and manufacturing techniques were increasingly informed by an awareness of the rapid transformation of Britain's own industrial technology. Indian tools and techniques of weaving or metal-working appeared crude, laborious and haphazard to Buchanan by the early 1800s, and indigenous textile production, far from being a continuing source of useful knowledge, rapidly came to exemplify the intrinsic backwardness of Indian technology.[17] It was several decades before the handloom industry, and related craft-based technologies, began to receive more sympathetic and constructive attention. By the end of the nineteenth century many village crafts – potting, weaving, basket-making, metal-working – had already been driven to near-extinction or were viewed by the colonial authorities as merely a cover for the activities of 'criminal tribes and castes'. Their trades and technologies were increasingly consigned, along with other 'relics of the past', to the ethnographical museum.[18]

---

[15] Roy (ed.), *Cloth and Commerce*; Peter Harnetty, '"Deindustrialization" Revisited: The Handloom Weavers of the Central Provinces of India, c. 1800–1947', *MAS*, 25 (1991), pp. 455–510.

[16] John Gascoigne, *Joseph Banks and the English Enlightenment: Useful Knowledge and Polite Culture* (Cambridge, 1994), p. 181.

[17] Dharampal, *Indian Science and Technology in the Eighteenth Century: Some Contemporary European Accounts* (Delhi, 1971), p. xxx.

[18] R. V. Russell and Hira Lal, *The Tribes and Castes of the Central Provinces*, IV (London, 1916); Desai, *Background*, p. 87.

MINES AND METALLURGY

Like textiles, mining and metal-working had for centuries underpinned the economic and cultural life of India. They, too, formed an important part of the early colonial reconnaissance of India's 'arts' and the subsequent critique of its apparently outmoded technology. The skill and inventiveness of India's craftsmen could not easily be denied: Delhi's Iron Pillar was alone testimony to the remarkable achievements of ancient ironsmiths. The production of *wootz* steel, used in swords and other weapons, and of a quality renowned in the East for centuries and admired in Europe well into the nineteenth century, was also the subject of much favourable comment and attempts at Western emulation.[19] However, here, too, there was growing disdain for Indian techniques – from the mining of minerals and metals to the quality of the finished product. The geologist Valentine Ball remarked in 1881 that Indian ironworkers had been criticised on the grounds that 'though they possess the art [of iron-making], they know nothing of the science'. But this, he noted, was not very different from artisans in 'more civilised countries'; even in European iron foundries, 'scientific guidance' was 'by no means universal'. Ball observed that, whatever its past attainments, iron and steel production in India in recent times had been adversely affected by the increasing scarcity of fuel for smelting, suggesting that not all the factors influencing the decline of India's old technologies were attributable to British rule. Even so, the amount of iron produced bore 'but a miserable proportion to the labour, time, and material expended': 14 tons of charcoal might be needed to produce a single ton of iron.[20]

Claims of inefficient and uneconomical production were commonplace throughout the nineteenth century, and mining and metallurgy had to wait several decades longer than the textile industry to experience the regenerating effects of modern technology. Although Indians had long mined iron, copper, gold, lead and other metals, the methods used, however well they may have served India in the past, were deemed crude and wasteful by contemporary European standards. Without gunpowder to aid mining operations, miners equipped with hand-tools could only hack out short galleries; without pit-props or pumps, they could seldom proceed very far before the shaft collapsed or filled with water, or the air became too acrid to breathe. Overcoming such problems in Britain's coal and copper mines had, conversely, been the nursery of its steam revolution. When copper mines in the Himalayan foothills were

---

[19] Dharampal, *Indian Science*, pp. l–liv.
[20] V. Ball, *A Manual of the Geology of India, Part III: Economic Geology* (Calcutta, 1881), pp. 338, 341.

examined by a Cornish miner in 1838 he suggested various improvements, including better tools, the use of timber props and steam-driven pumps. But few changes were introduced and the mines were abandoned in 1841, like so many others, as unremunerative; English copper was cheaper, even in the interior of India.[21]

The centuries-old Kolar gold mines in Mysore were rediscovered by the British in 1802, but were not considered viable for further working. Only after 1871 did new mining syndicates begin to rework the old shafts, and not until after the introduction of electric power in 1900 were mining operations carried to new depths.[22] Lead mining in Rajasthan had also once been highly productive, but it relied on primitive, labour-intensive techniques, the ore being pulverised with wooden clubs before being mixed with cow-dung for smelting. The resulting metal was said to be purer than that available in Britain, but it could not compete with the imported product even at Agra, being 'one-eighth dearer than English lead'.[23] Zinc mines in Rajasthan closed during the famine of 1812–13, presumably from a shortage of labour, and never reopened. Similarly, the mining of diamonds, one of the products that first established Indian opulence in Western eyes, was regarded in the nineteenth century as being carried out on an inefficient and largely unprofitable hit-or-miss basis, without the benefit of surveys and with only the most basic mining technology. Dry-season diamond workings at Panna in Bundelkhand were described in the 1860s as merely pits, seldom excavated to a depth of more than 50 feet before the water-level rose too high or the sides caved in. Earth was hauled to the surface in baskets, while Persian wheels, turned by bullock-power, struggled to extract water from the diggings. But few large and valuable stones were found and the profits were too small to attract European investment and technological innovation.[24]

Iron manufacture, once so widespread that Dharampal estimates that there were 10,000 iron and steel furnaces operating in eighteenth-century India, producing on average 20 tons of iron a year each, had been 'wholly crushed out of existence' by competition from Britain by 1880. From having been a major producer, India became one of the principal markets for Britain's iron (as for its cloth), with imports worth about £15 million a year.[25] In this case, some attempts were made to exploit Indian resources. The most celebrated example

[21] Ibid., pp. 267–8.
[22] R. D. Singh, 'Development of Mining Technology during the Nineteenth Century', *IJHS*, 17 (1982), pp. 208, 216.   [23] Ball, *Manual*, p. 300.
[24] Ibid., pp. 40–9; S. Bagchi and A. K. Ghose, 'History of Mining in India circa 1400–1800 and Technology Status', *IJHS*, 15 (1980), pp. 25–9.
[25] Dharampal, *Indian Science*, p. liv; Ball, *Manual*, pp. 338, 344–5.

was in south India, where in 1833 the European-owned Indian Steel, Iron and Chrome Company was established at Porto Novo on the Coromandel coast. For a few years it produced high-quality steel (taking ore from Salem); this sold well in Britain, but output soon suffered from a chronic shortage of charcoal. The company switched to producing pig iron, which required less fuel, but by the 1860s, with continuing shortages and high transport costs, this once-promising enterprise had collapsed.[26]

Coal mining, too, attracted British as well as Indian entrepreneurs. The Raniganj field 120 miles north of Calcutta was investigated as early as 1774 and the first excavations were dug in 1777. In 1836 the coalfield was purchased by the Bengali entrepreneur Dwarkanath Tagore, and, with railways in prospect, production doubled over the following ten years; by the 1880s there were more than forty mines scattered over an area of 500 square miles. By that date two-thirds of the coal consumed in India was produced in the country, with only a third imported, but the technology remained remarkably basic. As with other forms of mining, and as with later railway and canal construction, coal extraction relied on very labour-intensive methods. Tribal Santals and low-caste Bauri labourers, working in family groups of men, women and children, used picks and crowbars to prise coal from shallow pits and shafts. As late as 1886 the smaller, Indian-owned mines on the Raniganj field were said to look 'more like badly made tanks [reservoirs] than anything else'. In this industry, too, the introduction of electricity in the early twentieth century at last began to make the mines more efficient and allowed coal to be mined at greater depths.[27]

With the partial exception of coal, foreign competition, aided by the absence of tariff barriers and lack of technological innovation, held back the development of mining and metal-working technology in India until the early twentieth century. The relatively crude, labour-intensive nature of surviving mining techniques contributed to the false impression that India was poorly endowed with mineral resources or that they were inaccessible or otherwise difficult and unremunerative to work. But the fate of mining and metallurgy was affected by political as well as by economic and technological considerations. The British were aware of the part metal-working had played in supporting indigenous powers in the past through the production of arms and ammunition, and, just as they introduced an Arms Act in 1878 to restrict Indian access to firearms, so they sought to limit India's ability to mine and work metals that might sustain it in future wars and rebellions. This was especially

[26] Ball, *Manual*, pp. 349–50.
[27] Ibid., pp. 59–82; C. P. Simmons, 'Indigenous Enterprise in the Indian Coal Mining Industry, c. 1835–1939', *IESHR*, 13 (1976), p. 191; Singh, 'Development', pp. 212–15.

the case with Rajasthan, a region rich in metals. In the 1820s James Tod identified the 'mines of Mewar' as one of the means that had enabled its masters 'so long to struggle against superior power, and to raise those magnificent structures which would do honour to the most potent kingdoms of the west'. Indian skill in the difficult art of casting brass cannon had made Indian artillery a formidable adversary from the reign of Akbar to the Maratha and Sikh wars 300 years later. But by the early nineteenth century most of the mines in Rajasthan had been abandoned: the caste of miners was 'extinct', though Tod seemed unable or reticent to explain why.[28] During the Company period, as military opponents were eliminated and princely states extinguished, so was the local capacity to mine and work metals steadily eroded. As late as the Rebellion of 1857, the mining of lead for ammunition at Ajmer was perceived as a threat the British would no longer countenance and the mines were closed down.[29]

SHIPBUILDING AND STEAMSHIPS

Just as a capacity to mine metals and to convert them into coin and cannon had underpinned the military and economic power of India's pre-colonial rulers, so had the ability to build and maintain seaworthy ships been vital to the development of its overseas trade. Shipbuilding was a well-established craft at numerous points along the Indian coastline long before the arrival of the Europeans and was a significant factor in the high level of Indian maritime activity in the Indian Ocean region. Using local timber, especially teak from the forests of western India from Malabar to the Dangs of Gujarat, Indian ships were durable and well built for commercial use, though their design and construction were initially unequal to the challenge of the armed European merchant vessels that invaded Indian waters from 1498 onwards. There is some debate as to whether iron was used in Indian ships before the arrival of the Portuguese or whether iron nails and anchors were features subsequently adopted from Atlantic shipping.[30] As with cotton textiles, European trade was initially a stimulus to Indian shipbuilding: vessels built in ports like Masulipatam and Surat from Indian hardwoods by local craftsmen were cheaper and tougher than their European counterparts. While ships for local use continued in the main to follow traditional construction methods, those

---

[28] James Tod, *Annals and Antiquities of Rajasthan*, I (London, 1914), pp. 10, 399.
[29] Ball, *Manual*, p. 299.
[30] K. N. Chaudhuri, *Trade and Civilisation in the Indian Ocean: An Economic History from the Rise of Islam to 1750* (Cambridge, 1985), pp. 138–59.

built for European and some Indian customers for long-distance trade represented a successful marriage of Indian carpentry with European design. In addition to the use of teak, ships constructed for Europeans employed the technique of rabbeting by which the edges of planks were inset to fit closely together and form an almost seamless, watertight hull, thus dispensing with the European practice of caulking. Further Indian contributions were the use of coir ropes and the treatment of planks with lime to discourage wood-boring worms.

Between the seventeenth and early nineteenth centuries Indian shipyards produced a series of vessels incorporating these hybrid features. A large proportion of them were built in Bombay, where the Company had established a small shipyard. In 1736 Parsi carpenters were brought in from Surat to work there and, when their European supervisor died, one of the carpenters, Lowji Nuserwanji Wadia, was appointed Master Builder in his place. At first the Bombay yard performed only routine repair work, but in the second half of the eighteenth century ships of 500–800 tons displacement were being built there. In forty years as Master Builder, Wadia oversaw the construction of thirty-five ships, twenty-one of them for the Company. Following his death in 1774, his sons took charge of the shipyard and between them built a further thirty ships over the next sixteen years. The *Britannia*, a ship of 749 tons launched in 1778, so impressed the Court of Directors when it reached Britain that several new ships were commissioned from Bombay, some of which later passed into the hands of the Royal Navy. In all, between 1736 and 1821, 159 ships of over 100 tons were built at Bombay, including 15 of over 1,000 tons. Ships constructed at Bombay in its heyday were said to be 'vastly superior to anything built anywhere else in the world'.[31]

But, having reached a peak of production during the Napoleonic wars, the Bombay shipyard went into rapid decline after 1820. The last major vessel commissioned for the Royal Navy was launched in 1831, but even ten years earlier the dawning of the maritime steam age was beginning to impact on Bombay. The yard continued to have a twilight existence building timber hulls to house imported steam-engines: among these was the *Hugh Lindsay*, a 400 ton teak-framed steamer, launched in 1829, which pioneered the steamship route between Bombay and the Red Sea.[32] Steam vessels also began to appear on inland waters. The 132 ton *Diana*, launched from a dockyard at Kidderpore on the outskirts of Calcutta in 1823, worked as a harbour tug; a second vessel, the paddle-steamer *Pluto*, followed a year later. Like the railway locomotives in later

---

[31] Ruttonjee Ardeshir Wadia, *The Bombay Dockyard and the Wadia Master-Builders* (Bombay, 1955), p. 178.   [32] Headrick, *Tools*, pp. 135–7.

decades, few of these steamships and tugs were built in India. Most were assembled from prefabricated parts sent out from Britain, and so British engineering firms were technologically and financially the principal beneficiaries. Moreover, one steam technology obediently served another, as the many tons of iron rails and girders needed for railway construction began to arrive in Indian ports in the 1850s and 1860s in the holds of British steamships.

It could be argued that the decline of Indian shipbuilding was inevitable once iron hulls and steam-engines began to replace wooden sailing vessels; in other words, 'what destroyed Indian shipbuilding was British iron'.[33] But, Satpal Sangwan has argued, the demise of Indian shipbuilding owed more to political influence and economic self-interest than to the direct consequences of technological change. Like Indian textiles in an earlier age, the success of Indian shipbuilding alarmed British shipbuilders and shipping firms: pressure was applied through Parliament to restrict the entry of Indian vessels into British ports and to penalise their cargoes. The prospect of steamships being manufactured, not merely assembled, in India excited similar fears and so, impelled by British policy, India's shipbuilding industry was prevented from continuing to develop, even though it had a proven ability to adapt to changing technological needs.[34] The political point is worth making, but it is also necessary not to underestimate the extent of the technological gap between Britain and India at this time and the difficulty of Indian yards in matching the new capabilities and skills industrialisation had brought to British yards. It is also worth noting that, although work at the Bombay dockyards rapidly ebbed with the coming of steam, there were other lines of continuity. Shipbuilding was one of the routes by which Parsis moved from carpentry and other artisanal trades to become practitioners and patrons of modern science and technology. In his account of the 'master-builders' of Bombay, R. A. Wadia described the activities of Ardaseer Cursetjee, son of one of the last of the Parsi shipbuilders, who became an engineer and a pioneer of gas lighting and photography in Bombay, and who was Chief Engineer and Inspector of Machinery in the Bombay Dockyard until his retirement in 1857. He was also the only Indian in the nineteenth century to be elected a Fellow of the Royal Society of London.[35] Cursetjee's career and his wide scientific and technological interests provide a practical as well as symbolic link with the many other Parsis in the nineteenth and early twentieth centuries who became merchants

---

[33] Headrick, *Tentacles*, p. 367.
[34] Satpal Sangwan, 'The Sinking Ships: Colonial Policy and the Decline of Indian Shipping, 1735–1835', in Roy MacLeod and Deepak Kumar (eds.), *Technology and the Raj: Western Technology and Technical Transfers to India, 1700–1947* (New Delhi, 1995), pp. 137–52.   [35] Wadia, *Dockyard*, pp. 309–23.

and railway contractors, and later industrialists and scientists (among them the entrepreneur and pioneer of India's iron and steel industry, J. D. Tata, and the nuclear physicist, Homi Bhabha), and whose professional drive and private philanthropy contributed to the growth of hospitals, technical colleges, research institutes and other strategic sites of India's scientific modernity.

Britain's steam revolution was quick to display itself in India's capital cities and through the operations of the East India Company. As early as 1786 Matthew Boulton, one of the pioneers of the steam revolution, undertook the minting by steam-power in London of 100 tons of copper coins for the East India Company. Forty years on, Calcutta's new Mint, completed in 1831, was an epitome of how steam-power could now be harnessed to the service of the colonial state, its five imported engines driving rolling-mills, cutting presses, lathes and milling machines, and able to produce more than 300,000 coins a day.[36] Calcutta was witness to other early signs of the dawning age of steam, including India's first steam-propelled tugs. In 1825 a small steam-assisted paddle-steamer, the *Enterprise*, arrived at Calcutta, 113 days after leaving Falmouth, and three years later Lord Bentinck arrived on board the *Enterprise* to take up his duties as Governor-General of India. Once installed in office, Bentinck took up the cause of steam navigation on the Ganges. Following trials in 1828, iron-clad steamers began a regular service on the river, completing the 780 mile journey from Calcutta to Allahabad in three weeks instead of the three months previously taken by country boats.[37]

The speed and reliability of steam navigation were vital to the British in north India at a time when overland communications remained uncertain and slow, and a reliable mail service along the Ganges was a high priority for a government that had its headquarters far away in Calcutta. However, steam navigation had many limitations in India. The Ganges was not navigable by steam above Allahabad and barely beyond Mirzapur. The wide seasonal variations in the height and speed of India's rivers, along with their numerous sandbanks, shoals and other hazards, and the high costs of freight and passenger traffic, discouraged the more extensive development of this means of transport. In peninsular India there were hardly any rivers suited to steam navigation. Over much of northern India and Bengal, it was railways, not steamers, that ultimately supplanted the old river boats and ferries. Steam vessels were at their most effective on the Brahmaputra, where they facilitated the emergence

---

[36] Gascoigne, *Banks*, p. 229; Henry T. Bernstein, *Steamboats on the Ganges: An Exploration in the History of India's Modernization through Science and Technology* (Bombay, 1960), pp. 41–2.

[37] John Rosselli, *Lord William Bentinck: The Making of a Liberal Imperialist, 1774–1839* (London, 1974), pp. 285–92.

of the tea estates in Assam by importing 'coolie' labour from central India and exporting tea downstream to Calcutta. Steam vessels appeared on the Indus, too, in 1837, but their impact was limited. Steamers were expensive to buy and costly to run and in Sind there was no convenient source of cheap fuel. Above all, steamboats on the Indus were caught in a 'technological trap': vessels powerful enough to breast the river's strong currents were likely to be so heavy that they ran aground on shoals; if they were equipped with smaller, lighter engines they could not cope with the river's powerful flow. 'So the steamboats died a natural death', and greater reliance devolved on improved roads and railways instead.[38]

While steamboats lasted, their significance was arguably as much symbolic as practical: to many minds they were, and have remained, 'conspicuous heralds' of the 'approaching introduction of Western science and technology'.[39] Steam not only signified to Victorians the dynamism of their own civilisation; it also expressed the vast cultural and technological distance they saw between themselves and an India they perceived as backward and ridden by superstition. As one of the characters in William Arnold's novel *Oakfield* (1854) remarks to a fellow passenger on a Ganges steamer, there appeared to be an 'inconceivable separation ... between us few English, silently making a servant of the Ganges with our steam-engine and paddle-boats, and those Asiatics, with shouts and screams worshipping the same river'.[40] What Indian onlookers actually thought of steamboats, or many of the other innovations of the steam age, is not easily ascertained, but, to judge by the numbers of Indians who rapidly began to travel by steamboats and railways, or who greeted steam as an agent of their own modernity, it is unlikely that the cultural gulf was as wide as Arnold imagined.

ROADS, RAILWAYS AND THE POWER OF STEAM

Communication and transport by land, rather than on water, remained the greatest challenge of the age. There is an argument for seeing the paucity and poor quality of India's roads as both a cause and a consequence of its technological backwardness and so claiming that a revolution in transportation only came with the railways. 'The pace of transport [in India] had remained

---

[38] Clive Dewey, 'Some Consequences of Military Expenditure in British India: The Case of the Upper Sind Sagar Doab, 1849–1947', in Clive Dewey (ed.), *Arrested Development in India: The Historical Dimension* (Riverdale, MD, 1988), p. 138. [39] Bernstein, *Steamboats*, pp. 27–8.
[40] Francis G. Hutchins, *The Illusion of Permanence* (Princeton, NJ, 1967), p. 122; cf. Bernstein, *Steamboats*, pp. 50, 174–5; Michael Adas, *Machines as the Measure of Men: Science, Technology, and Ideologies of Western Dominance* (Ithaca, NY, 1989), pp. 172–3.

unchanged since remote antiquity', it has been said; with the coming of steam locomotion, it shot forward 'at an unimaginable rate'.[41] But such claims, for all their appealing melodrama, require some qualification. Land transportation was not as historically immobile as this extreme pre- and post-steam dichotomy would seem to suggest and the evolution of roads and road transport was important as both a prelude and an accompaniment to the growth of an effective railway system. The Mughals had built roads and bridges close to the centres of their own power or in the course of military campaigns, and some new forms of wheeled transport had emerged since the seventeenth century, without, however, reducing the dependence upon bullock-carts, donkeys and camels, or the pack-bullocks of itinerant Banjara traders.

Between 1780 and 1840, prompted by military considerations, the British began to repair, improve and extend India's roads. As with steamboats on the Ganges, securing and improving communications along the strategic axis between Calcutta and northern India was a priority for the Company, still locked in conflict with the Marathas and Sikhs. A military road cut from Calcutta to Benares in 1781 was followed in 1833 by the reconstruction of the Grand Trunk Road, using pounded *kankar* as a kind of Indian macadam to produce a hard, all-weather surface. By 1855 the road had reached Karnal, 75 miles north of Delhi, and soon thereafter traversed newly conquered Punjab en route to Peshawar. Other essentially military roads were built between the 1820s and 1850s: one, begun under Lord Wellesley as a military supply route and later converted into a metalled road, scaled the Bhor Ghat to connect Bombay with its hinterland in the Deccan, while in the southeast roads were built to consolidate the British hold on south India following the defeat of Tipu Sultan.[42] As this last event might remind us, in timing and function, as in its military agency, road-building paralleled and supplemented the work of the Trigonometrical Survey, helping physically and politically to impress the colonial presence on the Indian landscape. As well as strengthening military power, the Grand Trunk Road sparked a partial revolution in transport between Calcutta and the northwest. Horse-carriages began to replace slower modes of transport by the early 1850s. Lalla Tantimul, a contractor who had previously built boat bridges at Allahabad and Kanpur, pioneered a horse *dak* (mail and passenger) service in 1849; European competition soon followed, with the Calcutta-based North-West Dak Company soon operating over 1,200 miles of road. Gharries covered up to 100 miles a day, conveying passengers, by 156 stages, to Meerut. It had recently taken two and a half months by boat to reach

---

[41] Jean Deloche, *Transport and Communication in India Prior to Steam Locomotion, I: Land Transport* (Delhi, 1993), p. 1.   [42] Ibid., pp. 42–4, 71–2.

Benares; now letters from Calcutta arrived within 52 hours.[43] In Punjab, where there had been few roads suitable for wheeled vehicles before its annexation in 1849, road-building directly served military needs, but even there it also stimulated a sizeable growth in cart traffic as well.[44]

As often when their thoughts turned to colonisation and the technological tasks it presented, the British were reminded of the precedent of imperial Rome, whose roads were 'imprinted ... on the physical face of every country subjugated by the she-wolf, just as her laws and institutions were on the moral'.[45] And yet, even in the 1840s, after sixty years of sporadic road-making, the comparison with Rome did not flatter British achievement. The success of the Grand Trunk Road was largely unreplicated elsewhere, and there were few feeder roads to serve the main arterial routes. William Thornton's claim in the 1870s, that the British had found India 'as trackless as Britain was before the Roman invasion', but 'rendered it in most directions as permeable as England was in the early part of the Georgian era', unwittingly suggested the extent of the time-lag involved in road-building in India.[46] That famine relief workers replaced convicts as the main instruments of road construction in the second half of the century is indicative of the low-technology, labour-intensive mode of construction that continued to be used on the roads and a reminder of the role of famine policy in the creation of India's modern transport infrastructure. Roads continued to be built throughout the nineteenth century and, in areas such as the Central Provinces, where railways were slow to penetrate large tracts of country, they had a significant role in promoting economic integration and consolidating political authority.[47] Steam locomotion did not automatically obviate the need for older, more widely disseminated, transport technologies.

On the other hand, however, the inability of older modes of transportation to serve the needs of new technology was well illustrated by the difficulties Company officials encountered when they tried to exploit India's less accessible coalfields. In order to tap seams in the upper Narmada valley in western India it was decided in 1838 to experiment by sending 20 tons of coal overland from Hoshangabad to Bombay. For this purpose Banjaras were hired with their pack-bullocks, but they demanded a high price for their services and could not be dissuaded from flinging down the sacks of coal 'as they do grain'.

---

[43] 'The Grand Trunk Road: Its Localities', *CR*, 21 (1853), pp. 176–8.
[44] Dewey, 'Consequences', p. 122. [45] 'The Plains of the Lower Ganges', *CR*, 9 (1848), p. 15.
[46] William Thomas Thornton, *Indian Public Works* (London, 1875), p. 24.
[47] G. W. MacGeorge, *Ways and Works in India* (Westminster, 1894), pp. 65–104; Mahesh Rangarajan, *Fencing the Forest: Conservation and Ecological Change in India's Central Provinces, 1860–1914* (Delhi, 1996), pp. 46, 51.

An almost identical experiment was made ten years later. Again the Banjaras were uncooperative: they still flung down their loads and declared their bullocks 'would be destroyed by the pricking of the sharp points of the large pieces' of coal. This time, however, bullock teams lugged the coal as far as Dhari on the Narmada, where the sacks were transferred to country boats to continue the journey down-river to Broach. The boatmen, too, were unwilling to handle the unfamiliar and awkward cargo and several craft were lost or shed their loads on the rocks and rapids. It was concluded that the Narmada was 'not a navigable stream' for commercial purposes. After eight months, barely half of the consignment reached Bombay and at greater cost than imported Welsh coal. Trials on a steamship in Bombay harbour unsurprisingly showed Narmada coal to be of 'very poor quality'. A more significant conclusion was that India's inland coalfields could never be effectively and profitably exploited until they could be reached by rail.[48]

As in Britain, transporting coal was one of the main incentives behind the initial drive for railway construction in India, and, appropriately, one of the first lines to be completed, in 1855, ran 120 miles from Hooghly to the Raniganj coalfield. But pressure for railway construction came from other sources as well, including the increasing demand for raw cotton from the Deccan and northern India and the growth of India's export-orientated economy. The railways were intended to 'intermesh' the economies of Britain and India, giving British trade and industry greater access to Indian markets and to sources of cotton, oil-seeds, grain and other primary goods.[49] But railways, like irrigation canals, also occupied a more distinctively Indian niche, being partly intended as a technological solution to the famines that struck the North-Western Provinces and Madras in the 1830s and which recurred with devastating frequency throughout India between the 1860s and 1900s. In 1880, the Famine Commission initiated a new phase of railway construction, calling for a further 20,000 miles of railways to be built as 'protective', or anti-famine, works. At a time when other modes of state interventionism (including direct regulation of the grain market and price-fixing) were rejected as contrary to the doctrine of laissez-faire, railways were regarded as a legitimate state response, an effective means of combating famine while facilitating the desired growth of the market economy. Famine thus had an influential role in shaping the evolution and institutional forms of science and technology in India, not least by directing railway and canal development towards meeting

[48] E. Impey, *Memoir on the Physical Character of the Nerbudda River and Valley* (Bombay, 1855).
[49] Daniel Thorner, 'The Pattern of Railway Development in India', *Far Eastern Quarterly*, 14 (1955), p. 201.

perceived agrarian needs rather than through the development of an industrial base.[50]

But the colonial pursuit of railway technology was impelled, too, by the problem of India's size as well as its poverty. As the British Empire in India expanded (to embrace Sind in 1843, Punjab in 1849 and Awadh in 1856), and as new threats of rebellion or invasion arose, so the need for prompt and reliable communication between Calcutta and the rest of India, and between a proliferating number of provincial centres, correspondingly grew. In his minute on railway policy in April 1853, the Governor-General, Lord Dalhousie, stressed the 'immeasurable' political advantages to be derived from a system of internal communication that allowed 'full intelligence of every event' to be conveyed to the government more quickly than by any other available means. Railways would allow the government to mobilise its military forces more rapidly than before; but here speed was not the only gain, for soldiers also would be spared the long and gruelling cross-country marches that had hitherto magnified already alarming levels of troop mortality.[51] In effect, the engineer G. W. MacGeorge observed in the 1890s, railways had reduced India to a twentieth of its former size. Such was the 'power of steam' that places 400 miles apart were now in terms of journey-time only 20 miles away,[52] though a cynic might add by way of qualification that having three different railway gauges (broad, metre and narrow) was hardly the best means to reduce journey-time or fashion an integrated network.

There was an even more powerful ideological message attached to railways than there was to steamships. Locomotives and the railway network seemed to the British to furnish irrefutable proof of their material superiority and their commitment to 'civilising' and 'improving' India. In 1848, before railway construction had even begun, the Governor-General of the day, Lord Hardinge, told Sir Robert Peel, 'if we can proceed with our railways through the heart of the country, we shall make rapid strides in wealth and stability – for steam here would be the greatest instrument of civilisation for the people, and of strength for the Government'. On his own departure from India eight years later, Dalhousie, in a suitably mechanical metaphor, described the railways as one of the 'great engines of social improvement' his administration had brought to India.[53] A writer in 1847 had similarly remarked that it would be 'highly glorious to the British Crown and honourable to the British people, to enstamp on

---

[50] For the argument that railways did contribute to the reduction of famine, see Michelle Burge McAlpin, *Subject to Famine: Food Crises and Economic Change in Western India, 1860–1920* (Princeton, NJ, 1983), pp. 149–56, 174–6.

[51] M. N. Das, *Studies in the Economic and Social Development of Modern India, 1848–56* (Calcutta, 1959), pp. 71–2.   [52] MacGeorge, *Ways*, p. 221.   [53] Ibid., pp. 2, 103.

India and Asia this most characteristic symbol of the civilisation of the nineteenth century'.[54] MacGeorge, in his own paean in praise of steam in the 1890s, claimed it was 'one of the most remarkable triumphs of the steam locomotive to have aroused and awakened the Eastern world', to have 'undermined, and in great measure overturned, the larger number of those deep-seated ancient prejudices' – caste being not the least of them – 'which in India have so long and so tenaciously resisted all previous assaults'.[55]

Twentieth-century historians have been more divided in their assessment of the railways' socio-economic impact. For those who delight in the history of technological change, railways have an irresistible attraction. Their sheer scale seems to proclaim their importance. To cite some familiar statistics, by 1947 India could boast more than 40,000 miles of track, of which nearly 24,000 had been laid between 1853 and 1900. By the outbreak of the First World War India had the fifth-largest railway system in the world, the most extensive outside Europe and North America. By 1939 Indian railways carried more than 600 million passengers a year and 90 million tons of freight. Railways criss-crossed the subcontinent, drawing almost every part of India into their spider-web of steel. By 1900 there were few towns of any significance without a railway station, few cities that did not boast a large and imposing terminus, built like a castle or cathedral, and everywhere along the route, in bridges, tunnels, sidings and workshops, stood evidence of the immense engineering effort that had gone into building the railways of the Raj.[56] From the perspective of the enthusiast it is hard not to believe that the railways were pre-eminent among the principal agents of technological innovation and socio-economic change under British rule.

And yet it is only necessary to dip into the report of the Royal Commission on Agriculture as late as 1928 to read of the 'backward' state of India's railways, especially when the population and area served by India's railways are compared with those of Europe and North America, and to learn of the need for more roads to improve transport and communications in the countryside.[57] By the 1870s the railways had duly eclipsed the Banjaras' bullock teams in the long-distance haulage of salt and grain, but they did not displace the bullock-cart; indeed, the railways relied on country carts to bring raw cotton and other cash crops to the railheads or distribute grain to needy villages in times of famine. The railways were, moreover, 'a British show',[58] and, despite Karl

---

[54] 'Our Indian Railways', *CR*, 7 (1847), p. 369. [55] MacGeorge, *Ways*, p. 221.
[56] Headrick, *Tentacles*, pp. 53–7; L. S. S. O'Malley (ed.), *Modern India and the West* (London, 1941), pp. 238–42. [57] *Royal Commission on Agriculture in India: Report* (London, 1928), p. 369.
[58] Thorner, 'Development', p. 208.

Marx's confident prediction that railways would be 'the forerunner of modern industry' in India, most of the technological, as well as economic, benefits of the construction and operation of the railway system accrued to the world's first industrial nation. In the early decades, rails, sleepers and prefabricated bridges, along with the engines (and even their drivers), were imported from Britain, and though in the 1880s and 1890s Indian workshops produced small numbers of locomotives, pressure from Britain again thwarted the development of Indian competition. Between the 1850s and 1940s, more than 14,000 locomotives were sold to India compared with barely 700 made there; as with textile machinery, British engineering firms profited at the expense of India's productive capacity.[59] By 1914 extensive railway workshops had sprung up in places like Lahore, where there were 10,000 workers, but they remained largely confined to repair and assembly work.[60] Despite this, it has been argued in more positive vein that railway workshops (like that at Rawalpindi, where around 3,500 workers were employed) were 'ultra-modern factories', and they became 'schools of skills for managers and artisans' and provided 'a pool of specialist labour on which other enterprises drew'.[61]

Although there were no indigenous precedents for railroads and steam locomotives, it would be a mistake to see the railway system as a straightforward example of technology transfer from Britain to India. The construction of Indian railways, in particular, reveals inputs that reflect India's conditions and work practices. Some arose directly from the nature of the Indian environment. In the 1840s, before construction began, doubts were raised as to the practicality of railways in India because of its tropical climate and the attendant hazards of monsoon rains, floods, intense summer heat, rampant vegetation, and wild or unfenced animals. These arguments were rejected by the engineer E. W. Simms in his report to Parliament in 1846, and Dalhousie in 1853 was equally confident that there were no 'doubts and difficulties' that railway engineers and contractors could not overcome.[62] This reflected the unbridled technological confidence of the age, but it was more easily said than done. Although the railways were built by British engineers, and initially through British contractors, it was not possible simply to export British expertise to India. India's rivers presented engineers with unprecedented challenges, especially in attempting to span the broad river valleys of eastern and

---

[59] Fritz Lehmann, 'Great Britain and the Supply of Railway Locomotives to India: A Case Study of "Economic Imperialism"', *IESHR*, 2 (1965), pp. 297–306; 'Railway Workshops, Technology Transfer, and Skilled Labour Recruitment in Colonial India', *Journal of Historical Research*, 20 (1977), pp. 49–61.

[60] Ian J. Kerr, 'The Railway Workshops of Lahore and Their Employees, 1863–1930', in Surjit S. Dulai and Arthur Helweg (eds.), *Punjab in Perspective* (East Lansing, MI, 1991), pp. 67–77.

[61] Dewey, 'Consequences', p. 143.   [62] Das, *Development*, pp. 34–5, 73.

northern India, which were almost waterless, sandy wastes in the dry season but became raging torrents or broad sheets of flood-water in the wet season. Crossing them required exceptionally long bridges, built of iron and steel girders, with supporting masonry piers sunk deep into the river bed. The piles and caissons used in European bridge-building proved inadequate for the task and engineers learned from experience the value of the Indian technique (originally used in well-construction) of sinking weighted brick or masonry piers until they reached a stable depth.[63] By the time the East India Railway reached Delhi from Howrah in 1866 it had spanned the Son, Tons and Ganges (twice) with four massive bridges. That over the Son alone was seven-eighths of a mile in length, had twenty-eight spans of wrought-iron lattice girders, each 150 feet long, and cost £330,000.[64] Rivers were not the only challenge posed by the Indian environment. The steep ascent to hill-stations such as Darjeeling and Ootacamund, or the route from Bombay to Poona across the Bhor Ghat, also called for an enormous investment of labour and engineering skills. Requiring a long series of bridges, embankments and tunnels blasted through the basalt of the Deccan Traps, the 16 miles of the Bhor Ghat incline took seven years to build and, at its peak, in 1861, 42,000 workers were involved in its construction.[65]

The cultural characteristics of Indian labour were as influential as the physical environment. To some extent, railway engineers used the skills of artisan castes, like the renowned tank- and well-digging Wudders. But, in mobilising large numbers of unskilled labourers to construct the new lines, many British engineers and contractors were at first dismayed by the seemingly arcane work practices they encountered. In time, however, engineers and contractors came to accept them because, as Ian Kerr has argued, they were more concerned with getting the railways built as economically and expeditiously as possible than with the means by which they were built: the cheapness and relative abundance of Indian labour was worth many a machine. Under Indian conditions, the use of customary tools and techniques – in moving earth, digging wells, fashioning wood and metal – could be more efficacious than the use of seemingly more sophisticated implements and imported earthmoving equipment. Attempts to replace the 'rude tools' of the Indian earth-worker – mattocks, crowbars and baskets – with wheelbarrows or steam-driven digging equipment were largely unsuccessful, if only because they did not fit Indian ways of

[63] P. T. Cautley, 'On the Use of Wells, etc. in Foundations as Practised by the Natives of the Northern Doab', *JASB*, 8 (1839), pp. 327–40; MacGeorge, *Ways*, pp. 328–9.
[64] E. W. C. Sandes, *The Military Engineer in India*, II (Chatham, 1935), p. 119.
[65] MacGeorge, *Ways*, pp. 353–7.

working. This did not signify the absence of all innovation, and Kerr cites instances of Indian workers quickly and efficiently developing new skills, such as riveting. But alongside the carpenters, masons and well-diggers were armies of unskilled labourers, many working as family units. The large number of female and child labourers used provides one explanation for the size of the workforce but also for the persistence of labour-intensive methods of construction.[66] It took decades for railway engineers to surmount the challenges India posed, and not until the 1890s, forty years after construction began, had railway engineers sufficiently mastered the organisational, technical and environmental difficulties India presented for construction to become 'routinised'. Once acquired and tested, the experience laboriously gained by railway engineers in India was then passed on through professional journals and associations.[67]

TELEGRAPHS

For all its limitations and imperfections, the technology of large-scale construction and engineering works was one of the ways in which the colonial state came to stamp its authority on the Indian landscape and to reduce India's vast area and varied terrain to manageable proportions. To an even greater degree than the railways, the rapid spread of the telegraph represented the importance of the military and political impetus behind technological change in nineteenth-century India. Although not a direct consequence of steam-power, India's telegraphs, like those in many other parts of the world, progressed hand-in-hand with (and often in advance of) the railways. Part of the same technological complex, they often followed the same cross-country routes and, without telegraphic communication, railway signalling could be neither effective nor safe.

From a few miles of line in 1851, telegraphs had been extended over 4,250 miles of India and linked forty-six receiving stations by the end of 1856; they ran from Calcutta to Agra and the northwest, as well as connecting Bombay, Madras and Ootacamund. By 1865 there were 17,500 miles of telegraph lines, rising to 52,900 miles by the end of the century. By 1939 India's 100,000 miles of lines carried 17 million telegraphic messages a year. As with the railways, state involvement was a central factor but, apart from the period 1870–9 when railway construction was under direct government control, the state's main

---

[66] Ian J. Kerr, *Building the Railways of the Raj, 1850–1900* (Delhi, 1995), ch. 4.
[67] Ibid., p. 12; Ian Derbyshire, 'The Building of India's Railways: The Application of Western Technology in the Colonial Periphery, 1850–1920', in MacLeod and Kumar, *Technology*, pp. 179–83.

input was in providing guaranteed dividends to railway companies and in promoting the construction of lines needed for strategic purposes or as famine relief works. The telegraph, however, having arisen directly from political imperatives, remained a state concern.[68] The telegraph was a response to the need for a rapid and reliable system of information over the vast distances of the expanding Indian empire: one of its first uses was to carry news of the fall of Rangoon during the Second Anglo-Burmese War to Dalhousie in Calcutta in April 1852. It is oft-repeated exaggeration to claim that 'the electric telegraph saved India in the Mutiny of 1857', but the telegraph lines did demonstrate their importance by carrying to the authorities early intimations of the up-country revolt in May 1857 and, in its later stages, under the guidance of British military engineers, they followed close on the heels of Sir Colin Campbell's advance into an embattled Lucknow.[69] But the military did not command an exclusive use of a technology whose value was soon widely appreciated. The telegraph was rapidly adopted by Indian and European businessmen, and by the early twentieth century neither government officials nor nationalist politicians seemed able to function without a daily diet of telegrams. Once submarine cables were completed between India and Britain in 1870 (paralleling the recent opening of the Suez Canal to steamship navigation), the telegraph integrated India more completely than ever before into the administrative, military and commercial network of the British Empire.[70]

The rapid construction of the telegraph system in India in the 1850s also illustrates the continuing importance of the personal interest shown in science and technology by individual governors-general (and, from 1858, viceroys). Bentinck in the late 1820s helped advance the cause of the Trigonometric Survey and the use of steamships on inland waterways. At the turn of the century Curzon lent his personal authority to the development of 'imperial science' in such fields as agriculture and medical research. Dalhousie, another 'technophile', energetically pursued the twin technologies of railways and telegraphs (as well as introducing the less technology-linked postal service). Without his belief in their 'historical importance' (and the stimulus of the Mutiny and Rebellion) telegraphs would have waited much longer to 'take off' in India.[71] Dalhousie's technological zeal found a willing instrument in William O'Shaughnessy, one of the most versatile Company surgeons and professor of chemistry at Calcutta Medical College, who had conducted his own private

[68] Krishnalal Shridharani, *Story of the Indian Telegraphs: A Century of Progress* (New Delhi, 1956), p. 5.
[69] Sandes, *Engineer*, II, pp. 279–83; Deepak Kumar (ed.), *Science and the Raj, 1857–1905* (Delhi, 1995), p. 47.
[70] Sandes, *Engineer*, II, p. 294; Saroj Ghose, 'Commercial Needs and Military Necessities: The Telegraph in India', in MacLeod and Kumar, *Technology*, pp. 155–6.   [71] Das, *Development*, p. 140.

trials with electric telegraphs in 1839. Without having previously seen a telegraph system in operation, O'Shaughnessy built a 21 mile long experimental line near Calcutta, protecting it from the effects of tropical heat and humidity by using thicker cables than was customary in Europe and North America and covering them with an impervious layer of cloth and pitch. He showed that electrical signals could be sent over long distances in Indian conditions. His findings were not immediately taken up, but a decade later O'Shaughnessy was appointed by Dalhousie as India's first Director-General of Telegraphs. Before such posts were entirely Europeanised, O'Shaughnessy's principal assistant was an Indian, Shib Chandra Nandy, who was responsible for erecting the first line from Calcutta to Diamond Harbour and who oversaw the construction of lines to Allahabad, Benares and Dacca.[72] Although the telegraph largely represented imported technology, local expertise and the political patronage it received undoubtedly speeded its adoption and adaptation to local conditions.

IRRIGATION

Few technologies were more widely employed or held more importance for agrarian society and the state than those relating to water management. India's irrigation works displayed great diversity of form and function, ranging from temporary earth dams to the stone-built underground reservoirs and step-wells of Rajasthan and Gujarat and the inundation canals of northwestern India. In southeastern India, water storage and irrigation had long played a major part in enhancing agricultural production. Along the Tambrapani and Chittar rivers of Tinnevelly a series of *anicuts* or dams diverted water from streams into adjacent rice fields, and the surrounding countryside was dotted with tanks and wells.[73] Elsewhere in Tamilnadu, in 'dry zone' Salem and Coimbatore, river- and canal-fed irrigation was rare, but each district could boast up to 40,000 wells; many were sunk to such depths that teams of specially bred Kongu bullocks were needed to haul water to the surface. There is evidence to suggest, though, that many of the larger of these long-evolved systems of water storage and distribution were falling into decay by the early nineteenth century, whether as a result of warfare and political unrest, because incoming British officials were unaware of the need for regular maintenance, or because, under pressure to maximise revenue, they were unwilling to

---

[72] Shridharani, *Telegraphs*, pp. 12–13; Mel Gorman, 'Sir William O'Shaughnessy, Lord Dalhousie, and the Establishment of the Telegraph System in India', *Technology and Culture*, 12 (1971), pp. 581–601.
[73] David Ludden, 'Patronage and Irrigation in Tamil Nadu: A Long-Term View', *IESHR*, 16 (1979), pp. 347–65; cf. Nirmal Sengupta, 'The Indigenous Irrigation Organization of South Bihar', *IESHR*, 17 (1980), pp. 157–89.

allocate funds or mobilise labour for the purpose.[74] However, though during the course of the nineteenth century the British became preoccupied with large-scale irrigation works, the old irrigation works did not disappear or forfeit their ancient significance. On the contrary, wells and tanks survived in their multitudes, receiving fresh stimulus from the growth of cotton and other cash crops, or proving more reliable and less costly than irrigation canals. Provincial governments sporadically recognised the value of these 'minor works' and gave limited funds for their maintenance and repair. In 1901 private works (mostly tanks and wells) continued to supply water to nearly 60 per cent of land irrigated in British India; despite seventy years of public works, state schemes (mostly in the form of canals) accounted for barely 40 per cent.[75]

Although in the nineteenth century British military engineers devised irrigation works of unprecedented scale and design, their initial moves exhibited a significant degree of technological continuity. This was most evident in the deltas of southeastern India, where British works began as restoration schemes. In 1804 a survey of the Grand Anicut on the Cauvery, 'the greatest engineering work carried out in India before British rule began',[76] revealed that the northern branch of the river, known as the Coleroon, was receiving most of the flow while the southern branch, which supplied water to the rice-lands of the Tanjore delta, was drying up. Between 1835 and 1838 Captain Arthur Cotton of the Royal Engineers built a new anicut at the head of the Coleroon to direct water back into the southern channel. This was only partly successful but it marked the beginning of a series of weirs, bridges and channels constructed by the British over the next fifty years. By these measures the delta was saved from impending ruin while adding, inexpensively, a further 250,000 acres to the 750,000 already irrigated by the Cauvery.[77]

Cotton was conscious of building on earlier technology when he undertook a second scheme, to throw an anicut across the Godavari and thus to irrigate a dry and impoverished corner of coastal Andhra, severely affected by the famine of 1833. Begun in 1847, the first stage of the project, including a dam 12 feet high and 7,000 feet long, was completed in 1852, but the scheme was extended several times until by the 1890s it served a large part of the Godavari delta, enabling the intensive production of sugar, rice and oil-seeds.[78] Through

[74] David Hardiman, 'Well Irrigation in Gujarat: Systems of Use, Hierarchies of Control', *Economic and Political Weekly*, 20 June 1998, pp. 1533–44; Arun Bandyopadhyay, 'The Technology of Irrigation in a Colonial Agrarian Context: The Tamil Experience in the Early Nineteenth Century', in Kumar (ed.), *Science*, pp. 87–100.   [75] *Report of the Indian Irrigation Commission, 1901–03* (London, 1903), p. 11.
[76] Ibid., p. 9.
[77] Alfred Deakin, *Irrigated India: An Australian View of India and Ceylon* (London, 1893), pp. 157–8.
[78] G. N. Rao, 'Canal Irrigation and Agrarian Change in Colonial Andhra: A Study of Godavari District, c. 1850–1890', *IESHR*, 25 (1988), pp. 35–60.

his work Cotton became convinced of the incomparable virtues of canals – in providing water for irrigation, preventing famine, and giving India a cheap alternative to road and rail transportation. But, having worked on the restoration of the Cauvery irrigation works, Cotton discovered that the Grand Anicut – over 1,000 feet long and up to 40 feet wide – was built not of solid masonry but of earth and rubble encased in stone and cement. He believed that a similarly cheap but effective technique could be used on the Godavari and most of India's other rivers. The low costs of this 'rude mode of construction' helped persuade the Court of Directors to sanction the Godavari scheme. Although the masonry dams introduced by the British ultimately proved more durable and less prone to water seepage, in the short term the earth-and-rubble technique allowed Cotton, like later railway contractors, to produce works of modern engineering using large numbers of unskilled labourers equipped with only basic tools and with little recourse to skilled craftsmen and imported technology. Cotton might have a modern vision of how India could be improved by 'English energy and understanding' (backed by an evangelical conviction that the benefits of irrigation would help spread Christianity in India); but he freely acknowledged his debt to the 'old native engineers'.[79]

Irrigation schemes on the Himalayan-fed rivers of north and northwestern India presented very different conditions from the deltas of the southeast and owed rather less to indigenous precedents. They also called for greater technological ingenuity, but this was innovation in response to local conditions and under pressure of local necessity rather than in simple emulation of European techniques. The British found in northwestern India a number of short inundation canals, relying on the seasonal discharge of melt-water from the Himalayas, but which, without head-works to control the flow or adequate maintenance, had rapidly silted up. In 1817 Captain Blane was appointed to restore the Delhi branch canal constructed under the Mughal Shah Jehan. With this work completed in 1820, moves were then made to restore the old Jumna Canal near Delhi. Under the stimulus of the north Indian famine of 1837–8, these early works laid the basis for the far more ambitious Ganges Canal, begun in 1842 under the army engineer (and fossil-hunter) Proby Cautley.

Completed in 1854, the Ganges Canal was hailed by the Irrigation Commission in 1903 as 'a work which in magnitude and boldness of design has not yet been surpassed by any irrigation work in India or elsewhere'.[80] It was a purely European enterprise in the sense that it was preceded only by a

---

[79] Lady Hope, *General Sir Arthur Cotton: His Life and Work* (London, 1900), pp. 67–130; A. Cotton, *Public Works in India* (2nd edition, London, 1854), p. 55; R. P. Kulkarni, 'Irrigation Engineering in India', *IJHS*, 17 (1982), pp. 28–45.   [80] *Irrigation Commission*, p. 10.

short canal carrying water to the gardens of Meerut, but the technical difficulties to be overcome were largely unprecedented and presented particular problems to those, like Cautley, with little or no previous experience of hydraulic engineering. The canal required the erection of extensive head-works at Hardwar, where the Ganges entered the plains, to divert part of the river's flow into the canal, and the construction of huge masonry aqueducts to carry the canal across or under streams running down from the Himalayas, which became broad and surging torrents when the snow melted in summer. One of the largest and most imposing of these works was the Solani Aqueduct, 16 miles below Hardwar, its 750 foot waterway supported by fifteen arches, each 50 feet wide and resting on piers sunk 20 feet below ground level. The main work of excavation for the aqueduct was carried out by 'basket-labour', but wheelbarrows were introduced, a light tramway was installed, and in 1851 a steam locomotive was brought in to haul ballast, but it soon broke down and spent the rest of its working life driving machinery at the nearby Roorkee engineering college.[81]

Ultimately the Ganges Canal irrigated, via more than 5,000 miles of channels, 1 million acres of the Ganges–Jumna *doab*. To its admirers, the enterprise was unmatched anywhere in the world. 'Everything connected with the Ganges canals', declared Alfred Deakin in the 1890s, 'is upon a scale of magnificence' – from the head-works 'which battle with the river for miles', down to the 'masonry-lined channels, with drops and bridges of massive construction by which the surplus waters find their way back again . . . to the streams from which they came'.[82] Though engineers from India visited irrigation works in Egypt and Italy, these had little to offer: 'the Italian works', boasted one contemporary, 'though excellent in their way, appear positively insignificant when compared with the Anglo-Indian'.[83] The Ganges Canal provided its own engineering lessons, becoming in turn a model for hydraulic experts from the semi-arid regions of the United States and Australia. Soon after its completion, the Ganges Canal revealed a number of defects in design and construction. Extensive remodelling was needed between 1866 and 1870 to reduce the velocity of the water flow, to stop scouring from undermining bridge foundations, and to check regression of the canal bed. By the 1900s, through a series of further piecemeal improvements, the Ganges Canal had become far more efficient and less wasteful of water resources than when first opened.[84] In overcoming initial errors, the canal differed substantially from

---

[81] MacGeorge, *Ways*, p. 155.   [82] Deakin, *India*, p. 179.
[83] 'Italian Irrigation and Canal Legislation for India', *CR*, 21 (1853), p. 417.
[84] *Report of the Ganges Canal Committee* (Roorkee, 1866); Ian Stone, 'Canal Management and Agricultural Production in the United Provinces', in Dewey (ed.), *Development*, pp. 313–16.

Cautley's original design: first principles had been 'greatly altered, and shaped bit by bit to the circumstances of the country', 'amended from time to time until they represented the net result of the knowledge of irrigation gained in this part of India'.[85] Clearly, the Ganges Canal was a case neither of a technology transferred intact from the West nor of one substantially derived from indigenous precedents; rather it was an example of the kind of *'in situ* initiatives' developed by British engineers in response to local conditions.[86]

Cautley, like Cotton, was an army engineer. Despite the creation of an Indian Public Works Department in 1854, the great majority of engineers working on India's canals, as well as on many of its roads and bridges, on the Trigonometrical Survey and on some railways, continued till the end of the century to be army officers, mostly from the Royal or Bengal Engineers. This was not simply because few civilian engineers were available. As with the Indian Medical Service, the colonial regime saw the value of having an establishment of engineers employed on essentially civilian works but ready for war. 'What a Royal Engineer most needs', observed a government committee in 1888, 'is a knowledge of the resources of the country, the habit of dealing with the natives, and experience in understanding what they can and what they cannot be expected to do.'[87] Managing the land and managing labour went hand-in-hand. The army origins of India's engineers made it easy for them to believe in a kind of technocratic paternalism, which brought India the 'improving' benefits of railways, bridges and canals without the need for a more extensive engagement with Indian society. Deakin's phrase about canal works that 'battle with the river for miles' is suggestive of the manner in which army engineers saw themselves as conquering and subjugating an adversarial landscape, like the Roman forebears they so admired.

Annexation in 1849 opened the way for the extension of canal schemes into Punjab, the most militarised of India's main provinces and one of the areas where irrigation delivered its greatest economic, social and environmental impact. By the end of the century the construction of canals, like that of the railways, was still expanding rapidly under the pressure of famine and from the lure of enhanced state revenue from irrigated land. By 1892 there were nearly 43,800 miles of main canals and distributaries in British India, irrigating 13.4 million acres. By 1901 when the Indian Irrigation Commission set up by Curzon's government met to review the future of large-scale irrigation works, about a fifth of the total cropped area of British India was under irrigation,

---

[85] Deakin, *India*, p. 180.
[86] J. M. Powell, 'Enterprise and Dependency: Water Management in Australia', in Tom Griffiths and Libby Robin (eds.), *Ecology and Empire: Environmental History of Settler Societies* (Edinburgh, 1997), p. 118.
[87] Sandes, *Engineer*, II, p. 3.

though only a third by state schemes.[88] And yet, despite the sustained growth of the canal system and the sense of imperial achievement it generated, controversy continued to dog canal irrigation.[89] Until his death in 1899 Cotton criticised the Government of India for not building more canals and developing their navigational use. Others highlighted the high environmental and health costs of canal construction in north India – waterlogging, the contamination of once fertile soil with underground salts (*reh*), and the spread of malaria from land made swampy by leaky canals. As early as 1845 T. E. Dempster identified a close correlation between malaria and the advance of the West Jumna Canal. His concerns were enlarged upon in the 1860s and 1870s by critics such as A. F. Corbett of the Bengal Army, who declared that irrigation not only had failed to fulfil the extravagant claims made for it, but had become a source of new hardships, among which he counted the spread of malaria and the consequent 'emasculation' of canal-side populations. Famines, he claimed, had been more frequent and severe since canal construction began, the problem of drought had been exacerbated, and large tracts of land, 'covered with *reh*', had become unfit for cultivation. He concluded that there was 'very good reason to fear that there is not a single acre of land that has been irrigated by canals for ten years, whose produce has not very considerably diminished'.[90]

If Punjab seemed blessed by the transformation of its arid lands into irrigated wheat-lands, it appeared blighted in almost equal measure by malaria. The epidemic of 1908 that swept the province was acutely felt in the poorer parts of Amritsar, causing a 'holocaust' of sickness and death; but in seeking to explain the incidence of malaria in the town the medical establishment identified poverty, hunger and poor sanitation as being as much to blame as nearby irrigation canals.[91] The Irrigation Commission of 1901–3 hardly considered the environmental consequences of canals at all. Only with respect to the United Provinces (formerly the North-Western Provinces) was it noted in passing that canals might have 'injurious effects', but this was followed by the assurance that the realignment of canals and related measures had 'gone far to remedy, if they have not entirely removed, the evils of which there were such serious complaints in past years'.[92]

The claim that state irrigation works were a positive force for agricultural improvement and technological change has been strongly argued by Ian Stone

---

[88] *Irrigation Commission*, pp. 11–12.
[89] For the limited uses of canal irrigation for famine control, see B. M. Bhatia, *Famines in India* (3rd edition, Delhi, 1991), pp. 123–7, 196–200.
[90] A. F. Corbett, *The Climate and Resources of Upper India* (London, 1874), pp. 65–6.
[91] Clifford A. Gill, *Report on Malaria in Amritsar* (Lahore, 1917), p. 2.
[92] *Irrigation Commission*, Part II, p. 185

in his study of the North-Western Provinces. Far from seeing evidence of an environmental catastrophe of the kind Corbett discerned, Stone suggests that the negative effects of canal irrigation have been exaggerated and that, on the contrary, it fulfilled one of its principal objectives – to constitute a check against drought and famine. More than that, he believes, it stimulated production and a wide range of agricultural improvements: 'rational' peasants were not slow to seize upon a technology so readily available and well suited to their needs. Stone moreover argues that, although in some respects canals and wells were rivals, in reality they constituted 'complementary' or 'overlapping' systems: cultivators continued to find a use both for wells, especially in areas where canals were too distant or unreliable a source, and for irrigation canals.[93] But Stone's optimistic conclusions have been scorned by Elizabeth Whitcombe who, in line with nineteenth-century critics, has argued that the irrigation canals of northern India were 'a costly experiment'.[94] By the 1900s, however, the significance of canal irrigation was beginning to change, as the focus of state attention began to move away from the technical issue of water provision to the scientific improvement of Indian agriculture and as electric power and tube wells began to provide a more flexible and less geographically constricted means of supplying water for irrigation.[95]

TECHNOLOGY, IDEOLOGY AND RESISTANCE

For the British, large-scale construction schemes and engineering works such as the railways and irrigation canals were monuments to their power and munificence. In stone, steel and steam they embodied the idea of the British Raj as a technological empire, able by its grand works and feats of engineering to master forces of nature that had defied and enslaved Indians for centuries. Technology on such a scale was able (so it was believed) to eradicate famine, converting semi-arid wastes into shimmering fields of wheat and rice, and transform workshy peasants into model farmers. Confidence in technology as the engine of socio-economic change increased after the Rebellion of 1857, an event that seemed to make more direct forms of intervention in Indian society dangerously impolitic. Queen Victoria's proclamation of November 1858 promised 'to stimulate the peaceful industry of India' and to promote 'works

---

[93] Ian Stone, *Canal Irrigation in British India: Perspectives on Technological Change in a Peasant Economy* (Cambridge, 1984).

[94] Elizabeth Whitcombe, 'The Environmental Costs of Irrigation in British India: Waterlogging, Salinity, Malaria', in David Arnold and Ramachandra Guha (eds.), *Nature, Culture, Imperialism: Essays on the Environmental History of South Asia* (Delhi, 1995), pp. 237–59.

[95] Alfred Chatterton, *Agricultural and Industrial Problems in India* (Madras, 1903), pp. 3–64.

of public utility and improvement' in the hope that India's 'contentment' would be the best guarantee of Britain's 'security' – to which the rebel Begum of Awadh duly retorted that the Queen seemed to anticipate 'no better employment for Hindostanees than making roads and digging canals'.[96] A concern for security was reflected, too, in W. W. Hunter's claim in 1862 that when the railway system was completed it would increase the efficiency of British troops in India twenty-fold. In addition, 'the growth of commerce and internal wealth' brought about by the railways would by itself be 'a better security for the stability of our Indian Empire than all the legions of England and all her batteries of Lancaster guns'.[97]

In British eyes, technology articulated the widening intellectual and material gulf between Britons and Indians, and dwarfed whatever residual claims India might still make to possessing a civilisation of its own. The West's monopoly of steam-age technology seemed to extend almost indefinitely the period of India's tutelage and gave the British unprecedented confidence in the enduring impact of their rule. Deakin remarked in 1893 that, '[i]f the British in India had achieved nothing else, the public works policy of the past twenty-five years would fully justify their supremacy'. The canals alone were, in his view, a monument to British 'sagacity, ability, and magnanimity'.[98] A year later MacGeorge declared that no one acquainted with modern India 'would hesitate to assert that in the whole history of governments – not excluding that of ancient Rome – no alien ruling nation has ever stamped on the face of a country more enduring material monuments of its activity than England has done, and is doing, in her great Indian dependency'. Not only were these works 'stupendous' in their number and scale; they also excelled the 'magnificent Mohammedan or Hindu creations' of previous centuries in being 'engineering works of general usefulness which bear directly upon the material progress of the people'.[99]

This was not, however, a universal or unqualified opinion, and particularly in the aftermath of 1857 doubts were raised about whether modern Western science and technology were appropriate for India or in the best interests of its rulers. In seeking to explain the causes of the rebellion, the historian and former Company servant Sir John Kaye dwelt at length on how the railway and telegraph had 'disquieted' Hindu minds and provoked the scheming anger of the Brahmin priesthood, a 'privileged race of men, who had been held in veneration as the depositories of all human knowledge'. Unable to explain the

---

[96] S. A. A. Rizvi and M. L. Bhargava (eds.), *Freedom Struggle in Uttar Pradesh: Volume II, Awadh, 1857–59* (Lucknow, 1957), pp. 527, 531.
[97] Francis Henry Skrine, *Life of William Wilson Hunter* (London, 1901), p. 70.
[98] Deakin, *India*, pp. 10, 143; cf. Hutchins, *Illusion*, pp. 119–21.   [99] MacGeorge, *Ways*, pp. 1–3.

workings of the 'fire-carriage on the iron road' or messages sent at lightning speed by telegraph, they were shown to be 'as feeble and impotent as babes and sucklings'.[100] This view of Indians' primitivism and hostility to Western technological innovation was shared by many other European writers, and the cutting of telegraph lines and damaging of railway bridges and canal works during the course of the uprising were seen to give substance to the claim that a deep aversion to modern technology – from Enfield rifles to railway engines – had inspired the revolt or been one of the principal manifestations of its atavistic nature.

Such negative interpretations of Indian responses to modern technology deserve to be treated with caution and with due recognition of India's political plight and its cultural and social diversity. It is certainly possible to find instances where technological innovations were linked in popular perceptions and rumour with calamities such as drought and epidemic disease or with the more coercive and intrusive aspects of colonial rule. Railways might, with some justification, be implicated in the spread of famine and disease or the ruin of Indian handicrafts, but they also carried pilgrims in their thousands to ancient shrines and *melas* (festivals) and, like the train that bursts across the screen in Satyajit Ray's cinematic portrayal of Bengali village life, *Pather Panchali*, they epitomised the bustling energy of the unknown but alluring outside world. A recurrent motif in novels, memoirs and films, trains enabled Indians of all classes to embark on journeys in search of education, employment and self-discovery.

Medical and sanitary measures, from smallpox vaccination to attempts to regulate religious festivals or to examine travellers during epidemics of cholera and plague, constituted one of the areas where resistance was most likely to occur, if only because they were often seen to be physically threatening and culturally intrusive.[101] There were attacks, too, on survey parties, especially when they were seen to challenge local authority or presage increased taxation. Resistance, in turn, could be seized upon and exaggerated by British commentators as a way of projecting a stereotypical image of Indian ignorance and contrasting this with the rational, enlightened self-image of the European.[102] Acts of destruction undoubtedly occurred, as in 1857 with the cutting of telegraph wires and attacks on the Ganges Canal and Son railway bridge, but these can be explained in terms of the strategic interests of the mutineers and

---

[100] Sir John Kaye, *A History of the Sepoy War in India*, 1 (London, 1864), pp. 190–1.

[101] David Arnold, *Colonizing the Body: State Medicine and Epidemic Disease in Nineteenth-Century India* (Berkeley, CA, 1993), chs. 3–5.

[102] Matthew H. Edney, *Mapping an Empire: The Geographical Construction of British India, 1765–1843* (Chicago, 1997), pp. 326–31.

the intensity of rebel hostility to the colonial regime and the wider processes of economic and administrative change that accompanied it. Against these acts of destruction needs to be set the mutineers' own deadly use of Enfield rifles against the British in 1857, and the promise made by the King of Delhi, Bahadur Shah, in a rebel proclamation that Indian merchants would have 'the benefit of Government steam-vessels and steam-carriages *gratis*' once he was fully restored to power.[103]

The destruction in north India at a time of mutiny and rebellion further needs to be set against the enthusiasm with which Indians in Calcutta and Bombay had greeted the opening of the first railway lines a few years earlier and the alacrity with which they took up the use of steamboats, railways and telegraphs. In 1858, even before the final Mutiny campaigns had died away, one Calcutta-based commentator could claim that the railway was already 'quite naturalised in the native mind'.[104] Among the urban middle classes in particular, the technology the West cherished as its own was often seen as the common heritage of the modern age. In his 1860s' lectures, the Brahmo Samaj leader Keshab Chandra Sen reflected on the 'wonders' achieved by Western nations in 'modern times' and remarked on how, 'in the hands of a Christian Government', India had been 'adorned and blessed' with the benefits of new technology. 'In these days of civilisation and enlightenment', he told an audience in Calcutta in September 1866, 'of industry and enterprise, of trade and manufacture, of steam and electricity, of scientific discoveries and inventions, there are on all sides cheering indications of material improvement and prosperity.' But, as a moral and religious reformer, Sen was not overly impressed by these material advances. 'Modern civilisation' was 'eminently and essentially materialistic', he warned, and in the midst of all this 'pomp and splendour, ill fares the spirit'.[105]

STEAM, SWADESHI AND SWARAJ

As India in the late nineteenth century accumulated mills, workshops and factories, the question was again posed – were modern industry and technology right for India? The argument that India (and humanity as a whole) was better served by India's pre-industrial artistry and artisanal skill gained fresh support

---

[103] 'The Azimgarh Proclamation', in Ainslee T. Embree (ed.), *1857 in India: Mutiny or War of Independence?* (Boston, MA, 1963), p. 3.

[104] 'The East Indian Railway', *CR*, 31 (1858), p. 237; M. N. Das, 'Western Innovations and the Rising of 1857', *Bengal Past and Present*, 77 (1957), pp. 71–6.

[105] Keshab Chandra Sen, *Lectures and Tracts* (London, 1870), p. 49; *Lectures in India, Part II* (3rd edition, Calcutta, 1900), pp. 39–40.

from John Ruskin, William Morris and others who, in criticising the impact of industrialisation on Western society and its arts, looked with admiration at the 'almost inimitable' works of India and the East.[106] These views, in part wistfully romantic but potentially of great practical significance, were taken up in India from the 1870s by, among others, George Birdwood and E. B. Havell. As with Indian medicinal drugs, the organisation of exhibitions, in India and abroad, was an important stimulus to the rediscovery of India's arts and crafts. Having been called on to prepare a catalogue for the 1878 Paris Exhibition on the 'Master Handicrafts of India', George Birdwood of the Bombay Medical Service wrote glowingly of India's artisanal tradition. He claimed that in India everything, down to the cheapest toy or earthen vessel, was handmade and hence 'more or less a work of art'. In the West the machine age had brought untold 'social and moral evils', but in India traditional craftsmen still thrived, their skills honed to 'the highest perfection' by the 'training of countless generations'. Unlike their industrial counterparts in the West, Indian craftsmen had 'polluted no rivers, deformed no pleasing prospects, nor poisoned any air'; but sadly, Birdwood recorded, of late village craftsmen had been lured away in their thousands to the 'colossal' mills of Bombay, 'to drudge in gangs at manufacturing piece goods, in competition with Manchester, in the production of which they are no more intellectually and morally concerned than the grinder of a barrel organ in the tunes it evolves'.[107]

Birdwood's anti-industrial critique extended to agriculture, too. Ever since Charles Grant had penned his 'Observations on the State of Society among the Asiatic Subjects of Great Britain' in the 1790s, British ideologues and reformers had believed that there was 'prodigious' scope for 'improvement' in Indian agriculture, especially through the introduction of 'mechanical contrivances'.[108] But the record of Western technology in the Indian countryside was not overly encouraging. Some mechanical devices – cotton-gins, sugar-mills, maize-shellers and lightweight Swedish ploughs – had been adopted, but heavy steam-ploughs, like seed-drills and threshing machines, were unsuited to Indian soils and agricultural practices, and the experimental farms that tried to introduce them 'speedily became museums' for obsolete and rusting implements.[109] None the less, in the late 1880s, Birdwood rebuked Pandit Srilal ('a

---

[106] Thomas R. Metcalf, *An Imperial Vision: Indian Architecture and Britain's Raj* (London, 1989), ch. 5.
[107] George Birdwood, *The Industrial Arts of India*, II (London, 1880), pp. 131, 135–6.
[108] B. V. Subbarayappa, 'Western Science in India up to the End of the Nineteenth Century AD', in D. M. Bose, S. N. Sen and B. V. Subbarayappa (eds.), *A Concise History of Science in India* (New Delhi, 1971), p. 497.
[109] Albert Howard and Gabrielle L. C. Howard, *Wheat in India: Its Production, Varieties and Improvement* (Calcutta, 1910), p. 77.

distinguished student of the Royal Agricultural College at Cirencester') for his 'sweeping attack on the vernacular implements and operations of Indian agriculture' and embarked on a rapturous celebration of the 'Mahratta plough' and the rural life of Maharashtra. The 'perfected indigenous plough of the country', the 'product of three thousand years' experience', was contrasted with a lumbering steam-plough, which quickly sank into the 'soft, yielding soil' of the Deccan and could not be dragged out again. 'It had been recklessly brought into a sacrosanct economic system wherein it had no place.' It remained immobilised, 'bedaubed red, and worshipped as a *lingam*, the phallic symbol of Siva', a reminder that any attempt to bring steam technology to the Indian countryside was bound to be 'a flagitious and farcical failure'.[110]

Twenty years later, in 1901, E. B. Havell, Principal of the Government School of Art in Calcutta, also took a critical view of India's industrial future. He argued that it was wrong to assume that India could, or should, follow the West along the path of industrialisation. No friend of India, he averred, could view with unconcern the prospect of congested cities and depopulated districts, 'of unhealthy conditions of work, of struggles between capital and labour, uneven distribution of wealth, social unrest, and all the attendant evils of the great industrial development in Europe and America'. Besides, Havell doubted that the Indian artisan was fitted by 'disposition and habits' for such a contest: it would take generations for him to acquire 'not only the technical knowledge, but the business methods and business capacity necessary for success in an industrial struggle in European markets'. He believed that India's future lay instead in a revival of handloom production, giving its hard-pressed survivors the benefits of improved looms to boost their efficiency and productivity. Indians had taken to the sewing machine; could they not similarly adopt the fly-shuttle loom, and use it to produce textiles incorporating the finest traditional skills and designs?[111]

Havell's ideas had a wide impact. Missionary organisations such as the Basel Mission in Madras had already begun to popularise the fly-shuttle among weavers, and Havell's proposals were adopted by several provincial governments and 'progressive' princely states. In Bengal, P. C. Ghosh began to produce modified handlooms according to Havell's design. Some of these were displayed at the First Industrial Exhibition held in December 1901 in conjunction with the annual meeting of the Indian National Congress at Calcutta (and attended, significantly, by Gandhi while on a visit from South Africa). There was a growing debate among India's intelligentsia at the time about

[110] George Birdwood, *Sva* (London, 1915), pp. 56, 62, 75.
[111] E. B. Havell, 'The Industrial Development of India', *The Englishman*, 9 August 1901.

whether the country should try to develop new industries along Western lines or seek to capitalise on its own artisanal tradition by trying to revive and revitalise old crafts. This debate assumed fresh importance with the birth in 1905 of the Swadeshi movement, which combined protest against Curzon's Partition of Bengal with an attempt to develop India's industries and promote economic self-sufficiency.[112] As Havell reflected in 1907, the Swadeshi movement did not necessarily favour the revival of old crafts.[113] Indeed, many Indians spurned the anti-industrialism of Birdwood and Havell, believing this was an attempt to fob them off with a second-best and outmoded technology that would perpetuate India's economic dependence and further deny it access to the benefits and opportunities created by modern technology. They believed that India should pursue an industrial career in direct competition with Britain and other industrial powers. Textile factories, match works, steamship lines – all sprang up during the Swadeshi phase, though a large percentage of these undercapitalised, technically inefficient, ventures soon failed.

Ultimately, one of the enduring legacies of these years was the more extreme form of anti-industrialism represented by M. K. Gandhi in his 1909 tract *Hind Swaraj*. Gandhi believed that India had to reject the urbanism and industrialism of 'modern civilisation', which had been born in the West but was increasingly infiltrating India, deepening its economic and cultural dependence. Reacting to the poverty of famine-wracked India as much as to the political humiliation of colonial rule, Gandhi decried English education, the legal profession, medical practitioners and factories as all conspiring to 'deepen our slavery'. Deriding the very technology of which the British had been most proud, Gandhi remarked that railways, far from benefiting India, had spread plague, encouraged famine and strengthened the colonial stranglehold over India. It was, he wrote, 'beyond dispute that they propagate evil'.[114] Elaborating on the critique of industrial society made by Western writers such as Ruskin and Tolstoy, Gandhi denounced machinery, 'the chief symbol of modern civilisation', as a 'great sin'. Machines, which had begun to 'desolate Europe', had already 'impoverished' India: it was owing to Manchester that India's handicrafts had 'all but disappeared' and famine raged among the impoverished population. Morally and economically, India's salvation did not lie in industrialism. Workers in the mills of Bombay had become 'slaves'; women worked in conditions that were 'shocking'; and, if 'the machinery

---

[112] Dhruv Raina and S. Irfan Habib, 'The Unfolding of an Engagement: *The Dawn* on Science, Technical Education and Industrialization: India, 1896–1912', *SH*, 9 (1993), pp. 87–117.
[113] E. B. Havell, *Essays on Indian Art, Industry and Education* (Madras, 1910), pp. 56–65.
[114] M. K. Gandhi, *Hind Swaraj or Indian Home Rule* (Ahmedabad, 1939), pp. 41–2.

craze' continued, India would become 'an unhappy land'.[115] Gandhi advocated instead a return to the villages and the technology of hand-spinning and weaving. The *charka* and *khadi* (hand-spun, handwoven cloth) became, for Gandhi, the primary agents and symbols in India's quest for home rule (*swaraj*) and self-reliance. Gandhi's message – 'machinery is bad' – had little impact in 1909 but, after his rise to national leadership in 1918–20, his anti-industrialism and critique of modernity had a profound impact on the subsequent development of science and technology in India.

[115] Ibid., pp. 82–3.

CHAPTER 5

IMPERIAL SCIENCE AND THE INDIAN SCIENTIFIC COMMUNITY

Although there was no clear ideological or professional break between the science of the Company period and the science that followed under the Crown after 1858, there was a steady move away from the earlier modes of exploratory and observational science, in which the Company had taken an erratic interest, to a more confident alliance between science and the state. In the late nineteenth century the colonial regime employed science as both a means of self-legitimation and an aid to more effective government. While exploiting the authority and utility of Victorian science, the state remained none the less committed to a largely instrumentalist view of science: science existed to serve the empire, not to constitute an alternative source of authority or to dictate imperial priorities. However, from the 1890s through to the First World War there was unprecedented Indian interest in, and engagement with, Western science. With the growth of an Indian scientific community, India participated in international science in ways that belied any narrow definition of colonial science. The combination of these two elements – imperial science and an emergent Indian scientific community – did much to advance science in India in the critical decades of the 1890s–1900s to a position of intellectual and political prominence but also to fuel its inner tensions and contradictions.

SCIENCE AND THE SERVICES

Science played little part in the education and training of Indian Civil Service officers and, though a recreational interest in natural history often developed in the course of a career in India, there was always a suspicion of the professional scientist and a greater regard for the practical exercise of administrative authority. The ICS valued first-hand experience in the districts above the cosmopolitanism and intellectualism of science and regarded a close acquaintance with the villages and peoples of rural India and a grounding in the vernacular languages as a superior basis for knowing and ruling India. The Civil Service also jealously guarded its rights of social precedence and bureaucratic authority over the scientific and technical services and vigorously upheld the right to dictate to them on matters of organisation and policy. For as long as the scientific community was largely a European preserve this gave the ICS formidable powers of direction and control.[1] The state-run science that prevailed in India after 1858 was thus, to a large extent, the science of the 'enlightened' bureaucrat, egged on at times by the enthusiasm of a viceroy or provincial governor. This was a politically dependent science, and as such subject to financial and administrative pressures it was ill-equipped to resist.

After 1858 the state took up a number of projects pioneered by the scientific societies of Calcutta, Bombay and Madras or by zealous individuals. Among the most significant of these was the census. Although James Prinsep in Bengal and W. H. Sykes as Statistical Recorder to the Bombay government in the 1830s had conducted local demographic surveys, there was no systematic census until 1871. This launched a series of decennial censuses that embraced the whole of India, including the princely states, within a single 'imperial' formula.[2] More than a mere statistical record, the Indian census was a central feature in an expanding colonial governmentality, which, though it had its roots in the surveys of the Company period, marked the change to a more systematic and self-consciously 'scientific' regime of power under the Crown. Through essays on religion and ethnography, health and demographic change, as much as through its exercises in quantification, the census was a powerful force in articulating colonial ideas about India and (as we have seen in the case of malaria in chapter 3) in transmitting these to an Indian audience. Supplementing the censuses, and, from the standpoint of science, even more significant, were the imperial and district gazetteers. Arguing that nothing was 'more costly to a Government than ignorance', in 1869 the civil servant W. W. Hunter proposed a series of volumes that would provide a convenient digest of information about the Indian empire, its resources, inhabitants and administration. He saw the gazetteer as both an aid to better government and an enduring monument to British rule.[3] In 1871 the Government of India accepted Hunter's plan and the nine volumes of the first edition of the *Imperial Gazetteer* appeared in 1881. With additional data from the 1881 census, these were expanded into fourteen volumes in 1885–7, and further revised in 1907–9.

The *Imperial Gazetteer*, like the district gazetteers, aspired to be more than a compendium of place names and geographical descriptions. Its many volumes constituted a systematic and consciously scientific ordering of information

---

[1] Irene A Gilbert, 'The Indian Academic Profession: The Origins of a Tradition of Subordination', *Minerva*, 9 (1972), pp. 384–411; L. S. S. O'Malley, *The Indian Civil Service, 1601–1930* (London, 1931), pp. 114, 294–6.
[2] Bernard S. Cohn, 'The Census, Social Structure and Objectification in South Asia', in *An Anthropologist among the Historians and Other Essays* (Delhi, 1990), pp. 224–54.
[3] W. W. Hunter, 'Preface to the First Edition', *The Imperial Gazetteer of India*, 1 (2nd edition, London, 1885), pp. xxxi–xxxii.

about districts, provinces and an entire empire, and, in serving the needs of colonial bureaucrats, foreign observers and investors, they were based on a scrutinising externality that made India and its inhabitants minutely subject to the imperial gaze. It is significant that the gazetteers were planned and executed under the direction of an ICS officer and not a professional scientist. Hunter regarded the more detailed 'scientific aspects' of India as best left to more technical works. The gazetteers were confined 'to brief but careful sketches, such as might be useful to practical administrators'; the purely 'scientific inquirer' could look elsewhere for information.[4] But science was far from marginal to this imperial project, and the gazetteers display the strengths as well as the limitations of India's scientific empire. The introductory volume to the 1909 *Imperial Gazetteer*, for example, proceeds with positivist logic from the physical foundations of India, represented by its geology, meteorology, botany and zoology, through a survey of its human inhabitants, similarly compartmentalised into ethnology, language and religion, to conclude, as if with the highest stage in this evolutionary saga, with the departments of the colonial administration (public health and the police). Thus unfolds a grand panorama, viewed from the high ground of empire, in which each aspect of nature and culture had its assigned place, and in which the livelihoods and languages of the Indian people are neatly sandwiched between the zoology of fishes and the origins of the Aryan religion.

Significantly, only one of the ten introductory essays in the 1909 *Gazetteer* was entrusted to a scientist not directly employed by the Government of India (J. D. Hooker on botany); the rest were written either by members of the scientific and technical services in India or by ICS officers. The scientific ethos and methodology penetrate deep into the 'human' chapters: thus, the essay on ethnology and caste by H. H. Risley, ICS, opens with a technical discussion of craniometry and anthropometry as a 'test of race'.[5] Here, repeated in provincial and district volumes across the whole of British India, was an apparently comprehensive, uniform, up-to-date survey of an entire empire, embodying a view of India that was both external and scientific, but which also subordinated science to an ultimately political understanding of India and an administrator's sense of what constituted necessary knowledge. Given the bureaucratic nature of British rule (and the frequent changes in administrative personnel at district level), the gazetteers, along with the various other handbooks produced between the 1860s and 1920s, greatly contributed to the British sense of being able to 'know' India systematically and scientifically, and

---

[4] Ibid., p. xxx. [5] *The Imperial Gazetteer of India: The Indian Empire, I, Descriptive* (London, 1909).

from the authority of printed texts rather than the uncertainty of 'native' testimony. However illusory it might now appear, however imperfectly it equated power with knowledge, in the aftermath of the 1857 Rebellion this vast project constituted a kind of scientific title deed, proof to India and to the rest of the world of the secure understanding that Britain brought to its Indian domain.

If the *Imperial Gazetteer* represents the empire's view of science rather than the scientists' view of empire, this can be taken as a reflection of the enduring tension during the period of Crown rule, as earlier under the Company, between what scientists wanted to do and expected the state to do on their behalf and what the state was prepared to sanction and finance. In 1889, George Watt, the Government of India's Reporter on Economic Products, published a *Dictionary of the Economic Products of India*, a six-volume compendium of economic botany. Watt publicly announced the purpose of his *Dictionary* to be to 'supply scientific information which may be useful to the administrative officer', but also to advance the material interests of India and bring its resources more fully into the service of the empire.[6] Privately, however, Watt was indignant at the menial manner in which the government used his professional skills and the way in which scientific work was subjected to scrutiny and criticism by 'purely administrative officers' with no scientific training. He railed against the subordination of science to 'Secretariat interference', 'the curse of every Indian branch of special or scientific work'.[7] It was a complaint commonly voiced, if only in private, by India's scientific and technical officers. Watt hung on, but some, like Ronald Ross of the IMS in 1899, chose to resign in protest against their treatment and lack of research opportunities rather than endure further frustration and humiliation at the hands of the government or from their service chiefs. Similar conflicts between forestry officers and irrigation engineers on the one hand and their ICS masters on the other have recently been identified and confirm the view that science and technology were, with few exceptions, kept in strict subordination to the colonial civil service.[8]

There was, however, widespread recognition that in practice scientific and technical staff in India were bound by the nature of the imperial system to remain subordinate to the ICS. In his report on the state of Indian agriculture

---

[6] George Watt, *A Dictionary of the Economic Products of India*, 1 (Calcutta, 1889), p. vi.
[7] Deepak Kumar, *Science and the Raj, 1857–1905* (Delhi, 1995), p. 82.
[8] Ravi Rajan, 'Imperial Environmentalism or Environmental Imperialism? European Forestry, Colonial Foresters and the Agendas of Forest Management in British India, 1800–1900', in Richard H. Grove, Vinita Damodaran and Satpal Sangwan (eds.), *Nature and the Orient: The Environmental History of South and Southeast Asia* (Delhi, 1998), pp. 354–5; David Gilmartin, 'Scientific Empire and Imperial Science: Colonialism and Irrigation Technology in the Indus Basin', *JAS*, 53 (1994), pp. 1135–6.

in 1893, J. A. Voelcker of the Royal Agricultural Society pointed out the disadvantages, at a time when agricultural science had become a highly specialised field, of having a department of agriculture under an administrator with no scientific expertise. The early training of the civil servant was not 'one which directs his attention specially to, or encourages the pursuit of, natural science, but is rather one of a classical, mathematical, or literary character'. Despite this, it was assumed that a civil servant could be a successful jack-of-all-trades, as well able to run a department of agriculture as be a judge or district collector. The ICS officer who became Director of Agriculture might, moreover, serve only a short term before moving on to another, very different, post, and such technical expertise as he had acquired would again be lost to the department. Ideally, agriculture should have its own technical director, like the Geological Survey, and not a civilian; but, Voelcker conceded, many of the issues facing agriculture touched upon 'the circumstances both of the people and of the state', and hence some degree of civilian control over the department's technical staff was, from the government point of view, essential.[9]

Despite the tensions between the specialist scientist and the generalist administrator, despite complaints about Secretariat interference, poor pay and inadequate research facilities, many British scientists (and a score of Continental colleagues) in practice accepted that they were required to work within the political constraints of an imperial system and, like some of their predecessors under the Company, appreciated that India could give them unique scientific opportunities and a degree of public recognition scarcely attainable at home. Many scientists were encouraged, personally and professionally, by the fact that science as an aid to government was gaining strength in late nineteenth-century India, as the search for additional state income intensified and as the task of administration became more complex and technical.

Their practical achievements and impact on state policy were, however, often disappointingly meagre. In 1870 the Government of India appointed a medical man, Francis Day, as Inspector-General of Fisheries. His reports on freshwater and coastal fisheries demonstrated the practical benefits of developing a more scientific understanding of fish stocks, the breeding habits and habitats of different species, and fishing techniques in order to prevent further 'wasteful destruction'. At the same time, Day was critical of government policies, such as the impact of the salt tax on fish-curing and marketing, but to little effect, and when he retired the Fisheries Department was abolished.[10] Botany, too, enjoyed mixed fortunes. Although in the forefront of colonial

[9] J. A. Voelcker, *Report on the Improvement of Indian Agriculture* (London, 1895), pp. 302–5.
[10] Francis Day, *Report on the Sea Fish and Fisheries of India and Burma* (Calcutta, 1873), pp. 60–1.

science in the Company period, botany long continued in medicine's shadow and struggled to find state recognition of its own. Not until 1890 was the Government of India persuaded (through the efforts of Dr George King, Superintendent of the Calcutta Botanic Garden and Professor of Botany at the Medical College) to create a Botanical Survey of India. Its objectives were a characteristic mix of scientific research and economic utility. Its officers were to extend the exploration and recording of India's regional flora and coordinate the work of provincial botanists, but also to promote the introduction and cultivation of cinchona, tea, coffee and other commercial crops. Despite its diminutive size, the Survey proved one of the most effective of the state scientific services, not only in cataloguing provincial flora, but also for maintaining close contacts with Kew Gardens, the epicentre of imperial botany, of which King became Director in 1897.[11]

In the wake of a devastating cyclone that struck Calcutta in 1864 and the Orissa famine two years later, the government also turned its attention to meteorology. In 1867 H. F. Blanford was appointed Meteorological Reporter to the Government of Bengal; in 1875 the Government of India set up its own Meteorological Department. Headed by Blanford for the next fourteen years, this department was responsible for coordinating provincial reports, preparing daily weather charts, and recording seismic and solar activity. By 1878 India had more than 100 meteorological observatories.[12] Astronomy, too, received state assistance. N. R. Pogson at the Madras Observatory from 1861 to 1891 prepared a catalogue of 5,000 stars and discovered six minor planets. But it was his claim, presented to the Indian Famine Commission of 1878, that sunspots affected terrestrial weather conditions and so might influence the incidence of Indian droughts and famines, that elicited more committed state funding. An observatory founded at Kodaikanal in 1895 quickly gained an international reputation for the study of solar physics. It was brought under the Meteorological Department in 1899, thus extending that department's bureaucratic empire.[13] As the examples of meteorology and astronomy suggest, the government's awareness of the problems posed by, and to, the Indian environment, more especially the experience of famine from 1866 onwards, prompted the extension of its commitment to science in the late nineteenth century. And yet, as indicated in the previous chapter, the primary response was to look to railways, canals and, after 1880, the Indian Famine Code; science was only secondary.

[11] B. V. Subbarayappa, 'Western Science in India up to the End of the Nineteenth Century', in D. M. Bose, S. N. Sen and B. V. Subbarayappa (eds.), *A Concise History of Science in India* (New Delhi, 1971), pp. 519–20.   [12] Ibid., pp. 502–3.   [13] Ibid., pp. 505–6.

IMPERIAL SCIENCE

Although a proposal to establish a department of science under the Government of India was made in 1884, anticipating the need for central direction and pooled expertise for agriculture, botany, forestry and related fields, the idea was rejected, and almost twenty years elapsed before a similar scheme came into being.[14] The need more effectively to harness science to the state first became a matter of political urgency under the viceroyalties of Elgin (1894–9) and Curzon (1899–1905). In December 1898, at a time of growing popular unrest, especially over its policy towards the plague epidemic that had broken out two years earlier, Lord Elgin's government belatedly recognised its isolation from modern science, finding itself 'almost without competent advisers in many branches of science'. When scientific issues required investigation, responsibility fell on officials 'who are neither by experience nor knowledge competent to offer a decided opinion as to the best course to be pursued'. The result, Elgin concluded, 'has too often been misdirected action and the best endeavours have proved fruitless because the best men have not been secured for the work required'. The Government of India accordingly asked for the advice of 'leading men of science' in Britain to exercise 'a general control' over research work proposed or carried out in India.[15]

At a time when India was developing its own scientific expertise, not least through the endeavours of Indian scientists, this might be considered a retrograde step, one that showed a marked lack of confidence in the emergent scientific community in India and aligned colonial science even more firmly with metropolitan authority. From Elgin's request emerged the Indian Advisory Committee of the Royal Society, a group of scientific experts, some of whom had Indian experience. But the Committee remained inoperative until 1903, nearly three years after its formal constitution, and more immediate action was needed. In the wake of famine as well as plague, Curzon and his advisers decided in 1903 to create their own body of experts, called the Board of Scientific Advice, to bring together the heads of India's scientific and technical services in twice-yearly meetings, but, again deferring to metropolitan science, the Board was required to submit an annual report to the Royal Society in London. Curzon believed that in a rapidly changing world of science and technology there was an urgent need for sound advice and practical research in India, especially in view of the fact that the Government of India owned

[14] Kumar, *Science*, pp. 82–3.
[15] Roy M. MacLeod, 'Scientific Advice for British India: Imperial Perceptions and Administrative Goals, 1898–1923', *MAS*, 9 (1975), p. 351.

'the largest landed estate in the world, that the prosperity of the country is at present mainly dependent upon agriculture, that its economic and industrial resources have been very imperfectly explored, and that funds available for scientific research are limited'.[16] The relationship between the two advisory bodies was always fraught. The obligation on the Indian Board to report to the London Committee and the long delays in receiving replies accentuated resentment that only 'applied' and not 'fundamental' research was deemed suitable for scientific workers in India. There was annoyance that India should be subjected to carping criticism from experts in London, even if some of them were old India hands. The relationship in effect broke down in 1910: the Advisory Committee in London virtually ceased to function after 1914, a fate followed by the Indian Board in 1922.

Roy MacLeod has seen the career of the Indian Board and its struggles with London as evidence that a 'new sense of scientific independence was beginning to assert itself' in India.[17] While a desire for greater autonomy may have been in evidence in the 1900s, it was only after 1945 that 'scientific independence', like its political counterpart, was substantially attained. The continuing role of the Royal Society as the mentor and overseer of science in India and the fate of India's own Board of Scientific Advice might, in fact, be seen as indications of the long-term failure of scientists in India to wrest autonomy, or even a major share of the decision-making process, from the colonial state and its metropolitan allies.

While upholding the ICS view that scientific experts existed to advise, not to administer, Curzon was more alive than many of his bureaucrats to the scientific spirit of the age and to the practical, as well as polemical, needs of high imperialism. He clearly believed that the value of science was far greater than had hitherto been realised in India. Science (and not just the grand public works that had dominated nineteenth-century thinking) could be a force for far-reaching change, an aid to more efficient government, and not least, in an age of increasingly assertive nationalism, a fresh source of legitimation for British rule. Addressing a medical conference in 1899, Curzon asserted that the British had come to India not just as conquerors but also as benefactors, bearing the gifts of their law, religion, literature and science. There might be those who questioned the value of Britain's laws and religion, but about science, especially medical science, he said, there could be no doubt. Medicine alone was justification for British rule. It was 'built on the bed-rock of pure, irrefutable science'. It was 'a boon . . . offered to all, rich and poor, Hindu and

[16] Ibid., p. 355.   [17] Ibid., p. 371; Kumar, *Science*, pp. 105–9.

Mahommedan, woman and man'. Medicine lifted the veil of purdah 'without irreverence'; it broke down the barriers of caste 'without sacrilege'. Medical science was 'the most cosmopolitan of all sciences' because it embraced 'in its merciful appeal every suffering human being in the world'.[18]

Agriculture, so long neglected, was now a high priority. Laying the foundation stone in April 1905 for the Indian Agricultural Research Institute at Pusa in Bihar, Curzon stressed the Government of India's special responsibility for promoting scientific research and education. It had, he believed, a duty to 'give the lead' to the provinces by creating model scientific institutes and recruiting and training experts of the highest calibre. Central government was alone in a position to provide the necessary funding, direction and coordination. The achievements of its specialist institutes would be taken up in the provinces and the resulting scientific discoveries and educational benefits would be passed on, via local institutes and technical officers, to improve the 'position and prospects' of the peasantry.[19] Only the Government of India could recruit and suitably reward the European scientists Curzon believed India needed to revitalise its agriculture, the talented individuals who would lock India into the global scientific community at the highest level. Internally, Curzon saw India as a single scientific empire, not a set of semi-autonomous provinces, each pursuing its own parochial agenda with inadequate resources. For Curzon, science was essential to the modern, paternalistic state. He professed to have the well-being of the masses at heart and yet responded to the challenge of middle-class nationalism by urging India's princes and traditional elites to help fund medical and scientific research; when Indians proposed their own scientific and technical projects he was far less enthusiastic. The full credit for scientific and technological progress should rest with the imperial regime.

For Curzon the central government and the institutions and services answerable to it were vital to the formation of an informed science policy and the practical application of scientific research. In addition to a number of recently formed research institutes (veterinary science at Muktesar in 1898, agriculture at Pusa in 1905, forestry at Dehra Dun in 1906), by 1914 there were nearly a dozen all-India ('imperial') services and specialist departments, including medicine, meteorology, veterinary science, botany, agriculture, forests and geology. These varied widely in form and function: some – like the Geological Survey – were committed to both original research and its practical application, others were largely confined to data collection and surveys. Compared with the ICS and IMS, these imperial services were remarkably, even absurdly,

[18] *IMG*, 34 (1899), p. 134. [19] *Speeches by Lord Curzon of Kedleston*, IV (Calcutta, 1906), pp. 126–9.

Table 5.1 *Composition and pay of India's scientific services, 1920*[20]

	Europeans	Indians	European pay	Indian pay
			(Rs per month)	
Agricultural Service	38	5	1,000	460
Bacteriological Service	24	5	1,220	520
Botanical Survey	2	–	1,000	–
Educational Service	34	3	910	490
Forest Service	9	1	1,040	660
Geological Survey	16	–	1,010	–
Meteorological Department	10	2	970	770
Munitions Board	11	1	780	300
Trigonometrical Survey	46[a]	–	–	–
Veterinary Department	2	–	1,110	–
Zoological Survey	3	1	970	700
Total	195	18		

Notes:
[a] All except one Royal Engineers, holding military rank and rates of pay.

small, but even in 1920, when their numbers had been depleted by war service and casualties, the predominance of Europeans, and their superior conditions of employment, remained evident (table 5.1).

The vision of these services as representing 'imperial science' – a science that was India-wide and not merely provincial, devoted to serving the needs of India and the empire, carried out by scientists of world renown – was not confined to Curzon, but permeated the thinking of many scientists themselves. Despite the mounting pressure for Indianisation, these remained essentially European services, and their racial exclusiveness helped, despite the ever-present personal and professional conflicts, to shape a shared scientific culture and a common ideal of scientific service to the empire as a patriotic and paternalistic duty. That the state was by far the largest employer of scientists in India further strengthened a sense of cohesion, even idealism. **Albert Howard**, brought to Pusa as an economic botanist in 1906, was attracted to the post by the prospect of developing his research on tropical agriculture (begun in the West Indies) and by the manner in which, under Curzon, scientists had been recruited to serve India and bring the benefits of Western science to its people.[21] Even at the end of the First World War, the Indian Industrial

---

[20] Prafulla Chandra Ray, *Indian Science Congress: Presidential Address* (Calcutta, 1920), p. 16.
[21] Louise E. Howard, *Sir Albert Howard in India* (London, 1953), pp. 13–15.

Commission, under **Sir Thomas Holland**, former head of the Geological Survey of India, could favourably represent scientific and technical workers as a 'caste', with an *esprit de corps* of their own, and recommend the extension of the centralised service model to chemistry and other scientific fields.[22]

The converse of this service ethos was racial discrimination and what Deepak Kumar has called a system of 'apartheid' in the science of nineteenth-century India. He cites specific instances, including the attitude shown by the Superintendent of the Geological Survey of India, H. B. Medlicott, in 1880 in opposing the appointment of **Pramatha Nath Bose**. Medlicott remarked of Bose, recently qualified in London: 'he is a Bengali and may be physically unfit for our work.' Although Bose was admitted as the Survey's first Indian member, Medlicott remained hostile, declaring that his work as an assistant superintendent was only satisfactory when it followed an established path: 'when he afterwards moved to ground in which he had no outline to start with and the [geological] formations were new to him, his scientific helplessness became at once apparent.' Medlicott was convinced that Indians as a whole were 'utterly incapable of any original work in natural science'. Their recruitment should wait until 'the scientific chord' among Indians was touched, 'if indeed it exists as yet in this variety of the human race'. He urged the government to 'exercise a little discretion with our weaker brethren, and not expect them to run before they can walk'.[23] Bose suffered further rebuffs, eventually resigning in 1903 when Holland, ten years his junior, was appointed Director of the Geological Survey.[24] Bose's riposte, like that of a number of other Indian scientists, educationalists and administrators who found their talents snubbed in British India, was to seek employment in a princely state. As state geologist to the Maharaja of Mayurbhanj, Bose helped locate the iron-ore deposits that subsequently supplied the Tata iron and steel works at Jamshedpur, and so helped lay the basis for India's industrial economy.

The prejudice against Indians and complaints about their supposed physical and mental inadequacies extended to their employment as subordinates in the Geological Survey and other scientific departments. When in 1874 the Government of India directed that Indian 'apprentices' should be taken on by the Survey, it was at once decided to exclude Bengalis who 'from deficiency of stamina and manliness' were deemed incapable of the arduous field-work required. Several Sikhs, in colonial sociology a more 'manly' race, recent graduates from Lahore, were chosen instead, but Valentine Ball found little to praise

---

[22] *Indian Industrial Commission, 1916–18* (Calcutta, 1918), pp. 88–9.
[23] Deepak Kumar, 'Racial Discrimination and Science in Nineteenth-Century India', *IESHR*, 19 (1982), pp. 69–70.   [24] Kumar, *Science*, pp. 214–16.

in their work. In his view, Indians, 'however intellectual, and however well they may have been educated in various scientific branches, have shewn little capacity for undertaking original scientific research'. Indeed, they seemed to be 'deficient in synthetical powers to an extraordinary degree'.[25] Little wonder, then, that Indians filled less than 10 per cent of posts in the scientific services by 1920, or that the few who were admitted faced prejudice and discrimination.

Racial discrimination extended into the universities, where academic staff were 'practically segregated into two distinct racial camps'.[26] Indians, even the chemist **Prafulla Chandra Ray** with his Edinburgh doctorate, were confined to the provincial educational and scientific services, and not recruited (and only rarely promoted) into the European-dominated imperial services. The physicist **Jagadis Chandra Bose** was an exception in being directly admitted to the Indian Educational Service, but as an Indian probationer he received only a third the pay of his British colleagues until in 1903, after years of protest, he finally received full 'European' pay. Racial discrimination, it is sometimes suggested, prevented Indians from making a greater contribution to scientific research and intellectual leadership in the years after 1890. This is clearly true in a number of fields (medicine is a conspicuous example), but one could also argue the reverse – that the experience of racial discrimination made some Indian scientists, like Ray, even more determined to succeed. It should be borne in mind, too, that in their student days many Indians received encouragement and drew inspiration from European teachers (as, for instance, Ray did from Alexander Peddler, Professor of Chemistry at Presidency College). Nor were they always subject, even in India, to a crippling discrimination. Bose, in some respects, was relatively fortunate. In 1897, at the Viceroy's instigation, he was awarded a research grant of Rs 2,000 a year on the grounds that he was 'the first explorer and inventor in the electrical sciences that India has yet produced'. Such a grant was virtually unprecedented for either European or Indian members of the educational and scientific services, and the Home Department was sceptical that Bose was worthy of such an honour. 'It is difficult', a member of the Secretariat noted, 'to appraise or even to estimate, the services actually rendered since their nature is not such as to lend itself to an expression of results.' He conceded that Bose might be 'an original thinker (a rare thing in an Indian)' and acknowledged that 'his scientific work has earned him the highest encomiums from authorities who are best fitted to judge of it'. But, in a comment that revealed an antipathy to scientific research, not just its Indian practitioners, the Viceroy was reminded that salaries were

---

[25] V. Ball, *Jungle Life in India: Or the Journeys and Journals of an Indian Geologist* (London, 1880), p. 469.
[26] Patrick Geddes, *An Indian Pioneer of Science: The Life and Work of Sir Jagadis C. Bose* (London, 1920), p. 35.

'paid for definite services'; they could not be raised just 'because an officer is of high intellect'.[27]

MEDICAL SCIENCE

There were few scientific fields in which India seemed so deficient by the 1880s as medicine. At the time, Ross later recalled, the medical services in India were so poorly organised and resourced that medical libraries 'did not even exist in the principal stations'. Individual medical officers 'often did good clinical work', but there was little systematic investigation of specific diseases, 'and almost no establishment for it'. The great bacteriological discoveries of Pasteur and Koch 'were scarcely recognised, or were ridiculed', and Laveran's work on malaria, a disease whose control was vital to India, was 'almost unheard of'.[28]

Ross had, in retrospect, reason enough to be scathing, having felt that he was consistently obstructed by the government and the IMS chiefs in his own search for the malaria parasite in the early 1890s. But he was by no means alone in his negative assessment. When Ernest Hart, editor of the *British Medical Journal*, visited India in 1894 to address the newly instituted Indian Medical Congress, he was equally critical. Having reviewed the dramatic advances that had taken place in medical research over the previous fifteen years, and the way germ theory had revolutionised medical thinking, Hart asked why it was that all the major discoveries in tropical pathology had been made by foreigners – French, German, even Japanese – not by Britons. In an age of imperial rivalry, it was galling to have to recognise that the pioneering work on cholera, malaria and, most recently, plague had been done by others. In part Hart blamed the want of proper instruction in tropical medicine in Britain – a cause Patrick Manson was soon to take up – which meant that doctors arriving in the tropics had laboriously to acquire a knowledge of tropical pathology for themselves rather than being armed with it from the outset. But, no less substantially, he believed that research in India, far from being encouraged, was squeezed out by the burden of administration and ward duties. 'There is a feeling', he remarked, 'that a man with a leaning to science, with a new scientific fact in his head, is regarded by officialdom as a nuisance, as a sort of pestilential fellow with a new bug.' With promotion by seniority and not merit, it made sense to 'avoid giving trouble at headquarters' and to follow a course of 'respectable conservatism' rather than pursue potentially controversial research.[29]

[27] Gilbert, 'Profession', p. 401.   [28] Ronald Ross, *Memoirs* (London, 1923), p. 126.
[29] Ernest Hart, *The Medical Profession in India: Its Position and Its Work* (Calcutta, 1894), pp. 73–5.

The conservatism of the IMS and the lack of medical research can be overstated. Ross himself singled out exceptions – such as Henry Vandyke Carter, a pioneer of microscopy in India, whose work included studies of leprosy and relapsing fever, and T. R. Lewis and D. D. Cunningham, who investigated the aetiology of cholera. Although Lewis left India in 1883, Cunningham remained until 1897 as Professor of Physiology in Calcutta and ran one of the few research laboratories in India. And though there was resistance from some old India hands to the new science of bacteriology, symbolised by the scepticism that greeted Koch's discovery of the cholera bacillus, there were encouraging signs of impending change. Reflecting the needs of veterinary science, an Imperial Bacteriological Laboratory opened in Poona in 1890. The appointment of E. H. Hankin as Chemical Examiner for the North-Western Provinces in 1892 brought to India 'an original worker saturated with the bacteriological doctrines and teachings of Pasteur and Koch'. Hankin, who possessed a 'brilliant pen', did much to 'educate and stimulate' the medical profession in India out of its sluggishness.[30] He was one of the instigators of a meeting at Lahore in 1893 which pressed for the creation of an anti-rabies Pasteur Institute in India (eventually established at Kasauli in 1900). In 1893 the Russian-born, Paris-trained bacteriologist Waldemar Haffkine arrived in India to conduct trials with an anti-cholera serum. When plague broke out in 1896, Haffkine, the only full-time bacteriologist in India, developed an anti-plague serum at his small laboratory in Bombay. There were, then, signs of change by the mid-1890s, though significantly neither Hankin nor Haffkine belonged to the IMS or the Royal Army Medical Corps. Ross's momentous discovery of the role of the anopheles mosquito in the transmission of malaria showed that an IMS officer could produce research of international importance. But it was work done against the odds, owing much to the encouragement Ross received from Manson in London and virtually nothing to colleagues in India: 'We IMS men are not meant for research ... we are simply doctors,' Ross was once told by a colleague.[31] But it was indicative of the new mood that, when medical workers in India held their Congress in Calcutta in December 1894, the Viceroy, Elgin, attended and his government took seriously resolutions calling for the establishment of an all-India medical research institute and for central and provincial laboratories to investigate typhoid and cholera.

There were other, more urgent, pressures on the Government of India. The outbreak of bubonic plague in Bombay in 1896 brought a virulent new scourge to India: more than 400,000 deaths were recorded India-wide by the end of

---

[30] J. D. Graham, 'Medical and Research Organisation', *Seventh Congress of the Far Eastern Association of Tropical Medicine Souvenir: The Indian Empire* (Calcutta, 1927), p. 99.   [31] Ross, *Memoirs*, p. 243.

1900, and a further 8 million by the end of 1914. Like cholera before it, plague had international ramifications. For decades the British had been criticised at international sanitary conferences for not doing more to prevent the spread of cholera from India to neighbouring regions and, ultimately, to Europe and North America. Now in 1896–7 there seemed the likelihood that plague, too, would spread from Indian ports to the Middle East and Europe, unleashing a new Black Death. At the sanitary conference at Venice in February 1897 an international embargo was threatened against shipping from Indian ports unless the government took effective measures to bring plague under control. Fearing for the loss of its overseas trade, the government responded by introducing the Epidemic Diseases Act, one of the most draconian pieces of sanitary legislation ever adopted in colonial India. The Act authorised the compulsory hospitalisation of plague suspects, the destruction of houses and infected property, the physical examination of rail travellers, and the banning of fairs and pilgrimages. Although these measures helped calm international alarm, they provoked widespread resistance in the towns and cities of western and northern India over the following three years and deepened the government's dilemma as to how it should respond to the challenge of epidemic disease and mass mortality. Its response was further complicated when nineteen villagers at Malkowal in Punjab died in October 1902 of tetanus contracted from a contaminated needle while being inoculated with anti-plague serum from Haffkine's laboratory in Bombay. Haffkine was held directly responsible for the Malkowal 'disaster' and dismissed from his post in 1904; it was several years before his name was cleared and he never again had the opportunity to make a major contribution to bacteriological research in India. Contrary to rumour, anti-plague inoculation had not been made compulsory, but it was one of the most promising means of bringing the epidemic under control.

Eventually, in 1898–9, the government opted for a more pragmatic policy. It sought to avoid further confrontation by moderating its more extreme plague-control measures, accepting that large numbers of deaths were in consequence unavoidable, but trusting that public support for voluntary measures and an enhanced level of sanitary awareness would ultimately stem the epidemic. The Government of India grew wary of trusting medical and sanitary officers to make decisions that might have far-reaching implications for the relationship between the state and its subjects and was confirmed in its belief in the need for political control over medical policy.[32] None the less, the abandonment of the more coercive sanitary measures, the report of the Indian Plague Commission in 1901, and pressure from the medical establishment in

[32] Kumar, *Science*, p. 110.

Britain helped persuade the government that medical research had to be taken seriously.[33] The almost simultaneous discovery of the causes and modes of transmission of malaria and plague provided a much-needed boost to laboratory science in India, but also revitalised field research across a broad scientific spectrum, including medical entomology, helminthology, malaria surveys, and the investigation of the life-cycles and habitats of fleas, mosquitoes and other insect vectors.

Between 1900 and 1914 the government put in place the institutional structures and agencies that were to dominate medical research in India for decades to come, and which kept medical research in the state, rather than university, sector. A series of specialist institutions were created with the dual function of conducting research into plague, cholera and other tropical diseases and of producing the vaccines to combat them. These included the Central Research Institute at Kasauli in the Punjab hills in 1906, with special responsibility for malaria research, four state-aided Pasteur Institutes for the treatment of rabies and the preparation of anti-rabies serum, and two provincial bacteriological laboratories – at Bombay (renamed the Haffkine Institute in 1925) and at Guindy, Madras, where the King Institute of Preventive Medicine opened in 1904. To oversee the running of these institutes and to conduct research, a Bacteriological Department was created in 1906 (subsequently redesignated the Medical Research Department), with recruitment not confined to the IMS and an initial establishment of thirteen posts, soon raised to thirty. Through the initiative of Sir Harcourt Butler, Education Member of the Viceroy's Council, and Sir Pardy Lukis, the Surgeon-General, an Indian Research Fund Association (IRFA) was set up in 1911 to recruit and train medical researchers and to channel funds from the government and private benefactors into approved programmes of medical research. IRFA was a new kind of scientific agency for India (and even anticipated the creation of the Medical Research Council in Britain): it partly superseded the old IMS service model, but retained aspects of Curzon's focus on imperial science. IRFA symbolised the importance that scientific research had come to hold in state policy and in professional training and prestige. It was principally funded by an annual government grant of Rs 5 lakhs (the equivalent of £37,500) and at the height of its activities, in 1927–8, IRFA was spending Rs 11 lakhs a year on medical research in India. From 1913, research conducted under IRFA appeared in the *Indian Journal of Medical Research*, which rapidly established itself as leading international journal, publishing pioneering work on malaria, cholera, kala-azar,

[33] I. J. Catanach, 'Plague and the Tensions of Empire: India, 1896–1918', in David Arnold (ed.), *Imperial Medicine and Indigenous Societies* (Manchester, 1988), pp. 149–71.

plague and hookworm. Through the determined effort of Sir Leonard Rogers, IMS, Professor of Pathology at Calcutta Medical College, and with support from private funding, in 1921 the Calcutta School of Tropical Medicine was opened as a centre for the investigation of a wide range of tropical diseases.[34]

Taken together, these developments did much to reverse India's poor reputation for medical research and to cement its claim by the mid-1920s to have 'played a very distinguished part in the elucidation of tropical diseases'.[35] The research done, for instance, by Glen Liston on plague in Bombay, or by S. R. Christophers and a younger generation of IMS malariologists based at Kasauli, demonstrated that India was capable of frontline medical research. But the outcome was not all positive. The scientific work of the institutes and their small research staffs was often swamped by the demands of sera production, especially during the world wars when research came to a virtual standstill and energies were concentrated on producing vaccines for the armed forces. The dispersal of the leading research institutes throughout India made coordination difficult and undermined the Curzonian ideal of centralised research of all-India utility. Moreover, the old service ethos rapidly reasserted itself. Although it was recognised that, as outsiders to the colonial medical establishment, Hankin and Haffkine had helped get medical research moving in the 1890s, the Malkowal episode was not readily forgotten. It was held that, even though Haffkine had been a brilliant bacteriologist, he was a poor administrator and that, with few exceptions, IMS officers alone had the requisite combination of scientific skill and administrative authority to run a research institute. The IMS was denied a monopoly of research appointments, but it still retained a significant proportion of them. From research having been something the IMS did not do, the pendulum swung strongly in the opposite direction and the prospect of being able to conduct research on tropical diseases and possibly win a Nobel Prize (as Ross did in 1902 for his malaria work) was now advanced as a reason for the best medical minds to join the service. The 'facilities and opportunities' offered for medical research in India were unique, declared Lukis, now Director-General of the IMS, in 1917. 'There is scarcely one of the communicable diseases of India which will not repay further research.'[36]

The converse of this was the increasing isolation of medical research workers from everyday sanitary and medical practice. This elitism was exemplified not only by the professional and, still to a large extent, racial

---

[34] Helen Power, 'The Calcutta School of Tropical Medicine: Institutionalizing Medical Research in the Periphery', *MH*, 40 (1996), pp. 197–214.
[35] F. P. Mackie, 'Medical Research in India', in *Indian Empire*, p. 127.
[36] Pardy Lukis, 'Opportunities for Medical Research in India', *Indian Journal of Medical Research*, 4 (1917), p. 384.

exclusiveness of the IMS, but also by the fact that many of the medical research institutes and laboratories were located in the hills, in 'temperate' places, far from the main centres of population and from the diseases under investigation. Kasauli, at 6,000 feet above sea level, was a conspicuous example of this aloofness, but the fact that it was only three hours' drive from Simla, and hence under the watchful eye of the Government of India's Sanitary Commissioner, was seen as a strong argument in its favour. Other reasons were also advanced for keeping research laboratories and vaccine establishments in remote hill-stations. Public hostility to animal experiments and vivisection was one reason for moving the Imperial Bacteriological Laboratory from Poona to Muktesar, 7,000 feet up in the hills of Kumaon. Another argument used in favour of hill-stations was that the tropical climate of Calcutta and Bombay was inimical to advanced bacteriological work: serious science, it seemed, could be done only in cool locations and by research workers from the 'temperate' races. In 1891 the Viceroy, Lord Lansdowne, remarked to the Secretary of State that original scientific research demanded 'mental and physical qualifications' that were apparently not to be found in 'races bred in a tropical climate to the same extent that they exist in the more vigorous races of northern latitudes'.[37] Two years later the Principal of Grant Medical College reported that, whereas in Scotland the heart of a frog kept beating for at least 48 hours after it had been removed from its body, in the heat and humidity of Bombay it lasted barely 5 minutes. He concluded from this that sensitive physiological experiments could not be carried out except in the hills.[38] Such claims were taken seriously, with the result that medical researchers became a caste apart and medical research a privileged, predominantly white pursuit, deemed suitable for remote institutes but denied to under-funded university departments.

In fact, the isolation of many imperial research institutes bred boredom, bitter professional rivalries, and social and racial tensions, as well as provoking Indian criticism of the way in which science was being hidden from public scrutiny and accountability. When Britain's Surgeon-General was asked for his views on establishing a central bacteriological institute in the hills in 1900, he remarked that locating a small number of Europeans at a place like Muktesar, 'cut off from most of the advantages and comforts of civilised life', was bound to be 'very trying' for them. He cited the case of the forestry school at Dehra Dun to show that 'Europeans who remain for long periods in isolated places become especially liable to mental afflictions'.[39] In the short term, however,

---

[37] Kumar, *Science*, p. 191.
[38] G. A. Maconachie, to Surgeon-General, Bombay, 11 November 1893, in *Report of the Proceedings of the Central Indigenous Drugs Committee of India*, 1 (Calcutta, 1899), Appendix II, p. 7.
[39] W. R. Hooper, 10 May 1900, Revenue and Statistics Department, L/E/7/431: 1127, OIOC.

such remarks were ignored by an administration accustomed to believe in the virtues of hill-station life.

FORESTRY AND AGRICULTURE

Medicine was one area where the alliance of science and state grew appreciably closer during the late nineteenth and early twentieth centuries, but it was not the only one. State direction and a more consciously scientific approach also characterised colonial forestry from the 1860s onwards. In a work that underscores the importance of famine to the evolution of colonial science in India, Richard Grove has argued that the incentive for the development of colonial conservation and forestry policies came initially from an enlightened group of Company surgeons, who saw in rapid deforestation a major cause of desiccation, soil erosion and climatic change in India and hence a cause of India's apparently worsening droughts and famines. He claims that they were able to win government support for their scientific views, even though forest conservation ran counter to the Company's immediate commercial and financial interests.[40] Grove sees India more generally under the Company as playing a significant, pioneering role in the development of forest management and conservation policy, thereby demonstrating the error of those historians who have argued that 'science in the colonies was inherently secondary and far from the cutting edge of fundamental discovery'. In fact, he avers, 'the reverse seems to have been the case in many of the medical, field and meteorological sciences'.[41] There is, however, a danger of over-emphasising the impact of scientists (themselves mainly in government employment) on state policy, especially when scientific and technical officers were repeatedly reminded by their administrative superiors that their duty was to serve the empire, not to try to run it, and when financial and political considerations were so often given first priority. The extent to which surgeon-scientists were able to manipulate government policy under the Company thus remains in question. Madhav Gadgil and Ramachandra Guha have argued that in the second half of the nineteenth century the main incentive behind the government's forestry policy was a quest for financial gain rather than a simple acceptance of scientific arguments about desiccation and conservation. The setting up of an Indian Forestry Department in 1864, followed by the wide-ranging Indian Forest Acts of 1865 and 1878, can thus be seen as a pragmatic attempt to raise state income

---

[40] Richard H. Grove, *Green Imperialism: Colonial Expansion, Tropical Island Edens and the Origins of Environmentalism, 1600–1860* (Cambridge, 1995), ch. 8.
[41] Richard H. Grove, *Ecology, Climate and Empire: Colonialism and Global Environmental History, 1400–1940* (Cambridge, 1997), p. 2.

by tapping the relatively untouched wealth of India's forests as well as to meet the urgent practical need for a continuing supply of timber for railway construction, fuel and building purposes.[42] Deepak Kumar goes so far as to suggest that the Indian Forestry Service 'could scarcely be called scientific'; forest officers were 'basically estate-managers' and revenue 'was the only obsession'.[43]

This, too, may overstate the case and go too far in minimising the scientific significance of Indian forestry. Certainly, by the late nineteenth century forestry was one of the few government departments to make a profit, and in thus bringing science to the financial service of the state it served as an attractive model for many others. But state forestry is remarkable on other grounds, too. The size of its scientific domain was enormous, with 249,000 square miles under direct departmental control by 1916, and a further 99,205 square miles held as reserved forests.[44] Even more significant is the manner in which the 'scientific forestry' of the period demonstrated, despite undoubted inner tensions, the interdependence of science and state. Although Indian forestry was pioneered (as Grove has shown) by surgeon-botanists like Alexander Gibson and Hugh Cleghorn from the 1840s onwards, a more systematic and overtly scientific approach was introduced after the demise of the Company by Dietrich Brandis, appointed India's first Inspector-General of Forests in 1864. Born in Germany and trained as a botanist, Brandis held this office for nineteen years, during which time he propagated ideas of scientific forestry developed in Germany and France, and encouraged forestry officers to be trained in the doctrines and practices of the German and French forestry schools. To this extent he (along with other German-born foresters in the Indian service) was an important agent for the dissemination of metropolitan forestry.[45] But, despite government reluctance, Brandis was also instrumental in creating an Indian School of Forestry at Dehra Dun in 1878, which he saw as essential for training Indian assistants and for the more scientific management of Indian forests.[46] With his botanical background, he also strengthened belief in the utility of studying India's forests scientifically, publishing a *Forest Flora of North-West and Central India* in 1874 and a guide to *Indian Trees* in 1906, and overseeing the publication of other regional flora and specialist works on timber and forest products. As indicated in chapter 2, Brandis also advocated the use of Indian plant names, especially where they were more widely known and

---

[42] Madhav Gadgil and Ramachandra Guha, *This Fissured Land: An Ecological History of India* (Delhi, 1992), ch. 4. For a more qualified view, see Mahesh Rangarajan, *Fencing the Forest: Conservation and Ecological Change in India's Central Provinces, 1860–1914* (Delhi, 1996).   [43] Kumar, *Science*, p. 100.
[44] *Industrial Commission*, p. 39.   [45] Rajan, 'Environmentalism', pp. 343–4.
[46] Indra Munshi Saldanha, 'Colonialism and Professionalism: A German Forester in India', *Environment and History*, 2 (1996), pp. 195–219.

recognised than their English or Latin equivalents. Botany's greatest value for Brandis was as an applied sylvicultural science: 'good forest management' required a sound knowledge of individual trees (more difficult in an Indian than a European forest, because of the 'bewildering' variety of Indian species), along with their habits, modes and rates of reproduction, the climatic and soil conditions in which they flourished, and the properties of their timber. For Brandis, 'the study of the sylvicultural requirements of the different species' was the foundation 'of a successful system of treatment, leading up to the greatest annual production of timber and other forest produce per acre'. To him, as to his successors, the interests of science and state revenue were largely, and legitimately, indistinguishable, though he was at pains to point out that forests were a resource to be used for the ultimate benefit of the people and not just for short-term income-generation for the state.[47]

Inevitably, though, the establishment of effective state control over India's forests and the imposition of restrictions on their use for grazing, shifting cultivation and the collection of manure, fuel and building materials generated conflict with forest dwellers and those who lived on the forest margins and had long been accustomed to regard them as their own resource. From the tribal revolts of the 1870s and 1880s, through to the forest *satyagrahas* (Gandhian-style protest movements) of the 1920s and 1930s, forest regulations were a recurrent source of conflict, spilling over into the mainstream nationalist campaigns.[48] These conflicts could be understood as representing a fundamental and irreconcilable conflict between two very different systems of knowledge: on the one hand, the revenue-driven outlook of professional forestry officers, deriving their authority from the colonial state, Western science and ideas of woodland management originating in France and Germany; and on the other, the customary or local knowledge of indigenous hunters, food-gatherers and cultivators, who had an unschooled but deep and practical acquaintance with sylvan flora and fauna, which helped to provide them with their daily subsistence and with a trade in forest products. The distance between 'scientific forestry' and popular knowledge was often vast, and yet it would be excessive to see no meaningful engagement between the two. For instance, not only were forest-dwelling tribals employed as wage-labour to clear fire-breaks or cut

---

[47] Dietrich Brandis, *Indian Trees* (London, 1907; first published 1906), p. xxiv.
[48] For these conflicts see Richard H. Grove, 'Colonial Conservation, Ecological Hegemony and Popular Resistance: Towards a Global Synthesis', in John M. MacKenzie (ed.), *Imperialism and the Natural World* (Manchester, 1990), pp. 30–6; Atluri Murali, 'Whose Trees? Forest Practices and Local Communities in Andhra, 1600–1922', and Jacques Pouchepadass, 'British Attitudes towards Shifting Cultivation in Colonial South India: A Case Study of South Canara, 1800–1920', in David Arnold and Ramachandra Guha (eds.), *Nature, Culture, Imperialism: Essays on the Environmental History of South Asia* (Delhi, 1995), pp. 86–122, 123–51; Rangarajan, *Forest*, ch. 3.

bamboos; their knowledge was also used by forestry officers to identify species and inform their own understanding of forest ecology, or to locate and collect medicinal drugs, such as the *kino* gathered by Kurumbars in the forests of Malabar.[49] Over several decades, forestry officers also learnt from experience that once-derided practices, such as shifting cultivation, cattle grazing and the periodic burning of dried grass to stimulate new growth, might actually be necessary for the regeneration of sal, teak and other prized species.[50]

By contrast with forestry, and despite its evident importance for both subsistence needs and the export economy, India's agriculture was remarkably neglected as a site of scientific interest and activity before the 1890s. The tendency, as we have seen, was to look to large-scale public works to provide infrastructural solutions to India's agrarian problems and to combat famine. It is indicative of the special attention the state gave to forestry that a Forestry Service was set up in 1864 and a School of Forestry in 1878, more than thirty years before equivalent institutions were established for agriculture. Despite some provincial initiatives and some practical success in the dissemination of imported varieties of cotton and wheat, the general view was that agricultural improvement (as opposed to land revenue) was not a direct state responsibility but could be left to organisations such as the Agricultural and Horticultural Society in Calcutta, or that science had little to offer, compared with the grand technologies of irrigation canals and railways, to the progress and redemption of Indian agriculture. As indicated in the previous chapter, during the nineteenth century sporadic efforts were made to introduce European and North American agricultural machinery into India, but with meagre results. Against this trend, the Indian Industrial Commission in 1918 called for greater mechanisation of Indian agriculture, seeing this as a means both of increasing agricultural productivity and of stimulating industrial output.[51] The attempt, dating back to Kyd in the 1790s, to improve Indian agriculture by introducing new food plants also had a limited impact, and the efforts of Royle and his successors largely concentrated on tea, cotton, cinchona and other crops that were seen to have a clear commercial potential but were not necessarily attractive or suited to peasant agriculture. There was little systematic investigation of indigenous agricultural techniques, or of the properties and requirements of Indian staples such as rice, wheat and millet.

As with forestry, the development of agricultural science was greatly

---

[49] *Drugs Committee*, Appendix xxv, p. 169.
[50] Mahesh Rangarajan, 'Production, Desiccation and Forest Management in the Central Provinces, 1850–1930', in Grove et al., *Nature*, pp. 586–7; Ajay Skaria, 'Timber Conservancy, Desiccationism and Scientific Forestry: The Dangs, 1840s-1920s', ibid., pp. 596–635.   [51] *Industrial Commission*, ch. 5.

influenced by the problem of famine. In the wake of the 1866 famine in Orissa, Lord Mayo's government resolved to investigate and improve Indian agriculture, but the resulting Department of Revenue, Agriculture and Commerce lasted only until 1879 before being wound up for financial reasons. The famine of 1876–8 and the recommendations of the Famine Commission in 1880 produced a fresh surge of interest in 'agricultural improvement', with the central and provincial administrations forming their own departments of agriculture. But the Government of India's Department of Agriculture remained largely confined to an established routine of data collection and famine relief; it did little to apply science to agriculture. The main effect of provincial agricultural colleges was to 'turn out students with some knowledge of agriculture', who were then absorbed back into the revenue administration.[52] Proposals, first made in 1882, for the appointment of an agricultural chemist were repeatedly turned down and only finally approved in 1889. In his wide-ranging and influential report in 1893, Voelcker praised the government for having done so much for irrigation, thus meeting one of India's most fundamental needs, but was critical of it for neglecting modern scientific approaches to agriculture, especially the contribution chemistry could make to improving Indian soils, manuring practices and crop yields. As in medicine, Indian agriculture in the early 1890s seemed woefully out of touch with recent scientific developments.[53]

Famine was again the spur. In the famines of 1896–1900, Curzon was faced with further evidence of the catastrophic failure of Indian agriculture, and the report of the Indian Famine Commission of 1901 called the 'steady application to agricultural problems of expert research' 'the crying necessity of the time'.[54] Its recommendations were, for once, quickly acted upon. An Inspector-General of Agriculture was appointed in 1901 to oversee the scientific investigation of Indian agriculture and to supervise research in the provinces. A cryptogamic botanist followed, an entomologist was added in 1903, and in 1906 an Indian Agricultural Service finally came into being, forty-two years after the initiation of the Forestry Service. But the most important development was the creation of the Indian Agricultural Research Station at Pusa (later renamed the Imperial Agricultural Research Institute), made possible by a donation of £30,000 from Henry Phipps, an American philanthropist. This had a 1,300 acre experimental farm and a 'full staff of European specialists', including several of the agricultural experts appointed over the previous ten years, aided by 'native assistants'. By the time Pusa became fully

---

[52] F. G. Sly, 'The Department of Agriculture in India', *Agricultural Journal of India*, 1 (1906), p. 3.
[53] Voelcker, *Agriculture*, p. 314.
[54] *Royal Commission on Agriculture in India: Report* (London, 1928), p. 20.

functional in 1906, the imperial and provincial departments of agriculture together contained forty posts, including economic botanists, agro-bacteriologists and agricultural chemists. Only eight of these forty posts were held by Indians, and those mainly in subordinate positions, such as divisional inspectors; exceptionally, a Bengali, D. N. Mukherji, was the assistant director of agriculture in Bengal.

The articles that appeared in the early issues of Pusa's *Agricultural Journal of India* reflected the dominant lines of research: the effects of manuring on sugarcane, the improvement of paddy cultivation, insect pests in cotton, and the diseases of pigeon peas. Some of the most innovative work at the institute was carried out by Albert Howard and his wife Gabrielle Howard (appointed an economic botanist in her own right in 1913 – a rare instance of a woman occupying a scientific research post in colonial India). In 1893 Voelcker had argued that attempts to introduce English wheats into India had been based on the mistaken premise that Indian agricultural conditions were similar to England's: 'What is wanted is not so much to try exotic or imported seed, which may be good one year and fail to produce good results the next, but to try indigenous varieties which have already been found by experience of other districts to be well adapted.'[55] The Howards put this principle into effect, classifying and examining the properties of three dozen different varieties of indigenous wheat and from these developing rust-resistant 'Pusa' hybrids that were well suited to Indian conditions but superior in quality and market value to existing crops. Like Voelcker, the Howards saw the value of learning from peasant knowledge and not spurning it as most earlier experts had done. To try to reform Indian agriculture on Western lines, Albert and Gabrielle Howard argued in 1910, was a 'fundamental mistake'. The present agricultural practices of India were 'worthy of respect', however 'strange and primitive' they might at first appear to Westerners. What was needed was 'the application of Western scientific methods to the local conditions to improve Indian agriculture on its own lines'.[56] Through experimentation at Pusa and a second research station at Quetta, the Howards developed a holistic approach to the cultivation of wheat that took into account the life and welfare of the plant in relation to its environment, including such factors as manuring, soil conditions, irrigation, the effects of mixed planting and crop rotation, diseases and pests. Other important research was carried out on rice, a crop that had, despite its obvious importance to the diet and subsistence agriculture of millions of Indians, previously largely eluded agricultural science. The manner in which rice was able

[55] Voelcker, *Agriculture*, p. 240.
[56] Albert Howard and Gabrielle L. C. Howard, *Wheat in India: Its Production, Varieties and Improvement* (Calcutta, 1910), p. 7.

IMPERIAL SCIENCE AND THE INDIAN SCIENTIFIC COMMUNITY

to fix nitrogen and produce substantial yields on the same ground year after year was, in the long run, one of the most significant investigations conducted by Pusa scientists, including the agricultural chemist Jatindra Nath Sen.

But, like the hill-station medical research institutes, Pusa had its problems. Provincial governments were anxious to pursue research into how local conditions affected crops, or the development of local hybrids, and did not always take kindly to advice from the pundits of Pusa. Albert Howard, who left the institute after eighteen years, commented ruefully on the artificiality of life there and the personal and professional tensions rural isolation induced. There were few visitors and few social contacts except with other members of staff. In order to attract and retain European experts, they had to be paid high salaries, housed in palatial bungalows and provided, at considerable cost, such special facilities as schools, a hospital, a club (initially reserved for Europeans) and what were reputed to be the finest tennis courts in India. Without the stimulus of proximity to the wider scientific community, there was a steady loss of efficiency and sense of purpose. Researchers tended to 'lose a proper sense of proportion'; 'difficulties of all kinds' arose.[57] After buildings at Pusa were damaged by the Bihar earthquake of 1934 it was decided to transfer the Agricultural Institute to Delhi. It is hard not to see the abandonment of Pusa and the move to India's new capital as symbolic of the wider fate of imperial science and the secluded institutes it had spawned.

AN INDIAN SCIENTIFIC COMMUNITY

The scientific developments of the period 1890–1914 cannot be explained by reference to imperial science alone. It is also necessary to take into account the advent of an Indian scientific community, which became conspicuous in the 1890s but built on earlier trends. One long-term factor in its emergence was the growth of Western education in India, supplemented by Indian access to higher education and professional training in the West.[58] Calcutta's elite educational institutions were important both for the interest in science they stimulated and for the opportunities they gave for individual advancement. **Mahendralal Sircar**, physician and founder of the Indian Association for the Cultivation of Science, the physicist Jagadis Chandra Bose and chemist Prafulla Chandra Ray all attended the prestigious Hare School in Calcutta. Mahendralal went on to Hindu College before joining Calcutta Medical College in 1855 and qualifying with an M.D. in 1863. The geologist Pramatha

[57] Howard, *Howard*, p. 25.
[58] For a discussion of scientific and technical education in India in this period, see Kumar, *Science*, ch. 4.

Nath Bose was educated at Krishnagar College before joining St Xavier's in Calcutta, winning a Gilchrist scholarship to study in London in 1874. Jagadis Chandra Bose also attended St Xavier's, where his interest in physical science was stimulated by the teaching of a Jesuit, Father Lafont, before going on to study in Cambridge and London. Ray, through ill health largely self-educated, attended Presidency College until he, too, received a Gilchrist scholarship, which enabled him to travel to Edinburgh to study chemistry in 1882.

But too much significance should not be attached to Western education alone. The first generation of Indian scientists were also heirs to intellectual and scientific traditions of their own. The Bengali *bhadralok* drew intellectual sustenance and cultural pride from a long tradition of indigenous learning, backed by at least a modicum of wealth, and high social status (coming principally from the Brahmin, Kayastha and Vaidya castes). Jagadis Chandra Bose, like Pramatha Nath Bose, from the Kayastha or writer caste, was born into a relatively prosperous family in Vikrampur, an ancient centre of Sanskrit and Buddhist learning in east Bengal. His Brahmo father, a deputy magistrate and later an assistant commissioner, dissuaded Bose from joining the ICS and encouraged him to pursue an alternative career. Ray also came from a landed, literate background in east Bengal: his father was a Persian scholar and one-time student at Krishnagar College. Ray, too, was drawn to the Brahmo Samaj, which he joined in 1882. The intellectual excitement of student life in Calcutta opened up for him a lifelong enthusiasm for Western literature as well as science, though this was always tempered by the kind of patriotic spirit that led him to write a prize-winning essay on the Indian Mutiny while a student at Edinburgh. Mahendralal Sircar, by contrast, came from the relatively low-ranking Sadgop community and claimed to be a peasant's son, though support from his uncle, a Calcutta lawyer, saw him through medical college and into what was a relatively lucrative medical practice until his dramatic conversion to homoeopathy. Beyond Bengal, a not dissimilar melding of social background and educational opportunity was also discernible. The physicist **Chandrasekhara Venkata Raman** was born into a Tamil Brahmin family: his mother came from a line of Sanskrit pandits; his father, from a family of minor landholders, became a physics lecturer at Vizagapatam College. Despite a B.Sc. and M.A. in mathematics from Madras University, Raman was at first unable to pursue a scientific career and in 1907 joined the Indian Finance Department. He was posted, providentially, to Calcutta, the capital of Indian science.[59]

[59] Geddes, *Pioneer*; Jogesh Chandra Bagal, *Pramatha Nath Bose* (New Delhi, 1955); Prafulla Chandra Ray, *Essays and Discourses* (Madras, 1918); Prafulla Chandra Ray, *Life and Experiences of a Bengali Chemist*, I (Calcutta, 1932); Sarat Chandra Ghose, *Life of Dr Mahendra Lal Sircar* (2nd edition, Calcutta, 1935); L. F. Rushbrook-Williams (ed.), *Great Men of India* (London, 1939).

Sircar and Raman, like the self-taught mathematician Srinivasa Ramanujan, were somewhat exceptional among the leading figures in Indian science in this period in not receiving part of their education abroad. Until the 1900s there were few opportunities to take higher degrees and pursue research in Indian universities. Travelling abroad, usually to Britain, gave Indian students not only a superior training to any they could receive at home, but also the sense, hard to attain in India at the time, of belonging to an international scientific community – being part of the latest research activity, mixing on more equal terms than was ever likely in India with leading scientists, gaining access to a scientific domain that was not bounded by Britain alone but embraced France, Germany and the United States as well. Pramatha Nath Bose later recalled the exhilaration he felt when his first research paper was published in a science journal in London: 'visions of occupying a niche in the modern Temple of Science began to float before my mind.'[60] The six years Ray spent at Edinburgh working with P. G. Tait and Alexander Crum Brown, two of the principal figures in physics and chemistry at the time, nurtured his enthusiasm for scientific research, but also made him acutely aware of India's deficiency in producing scientists of its own. As a student, he later recalled, 'I found to my regret that every civilised country including Japan was adding to the world's stock of knowledge but that unhappy India was lagging behind'. He dreamt of a time when India, too, 'would contribute her quota'.[61] For Ray, as for P. N. Bose, the experience of returning to India to encounter racial discrimination, European domination of the services and grudging official recognition for his scientific qualifications was made even more galling by having previously known, and participated in, a more open scientific community. But the experience also made Ray and others even more determined to try to transcend, through a dual dedication to science and nation, the prejudices and pettinesses of the colonial world.

Some insight into the scientific world that J. C. Bose and P. C. Ray entered on their return to India can be gleaned by looking at the Asiatic Society of Bengal, still in the 1890s India's principal scientific forum. In March 1895, when Bose was elected to the Society, it had six subject committees: philology, numismatics, history and archaeology, natural history, physical science, and anthropology. The membership of these committees included some of the leading names in Indian science at the time, including P. N. Bose, D. D. Cunningham, T. H. Holland, G. King, D. Prain and R. D. Oldham. Apart from 22 honorary members (mostly eminent European scientists who had never set foot in India), the Society had 295 ordinary members at the close of 1894. But when Bose rose at the monthly meeting in May 1895 to present the paper on

[60] Pramatha Nath Bose, *National Education and Modern Progress* (Calcutta, 1921), p. 52.
[61] Ray, *Life*, p. v.

electric waves that launched his international career, only thirteen members and three visitors were present. Seven months later when Ray gave his paper on mercurous nitrite at the December meeting, he was the third of four speakers, sandwiched between Grierson on Kashmiri and Prain on Indian poppies, but at least that evening the audience touched twenty.[62]

The low turnout was not perhaps surprising. Of the Society's 295 members, only a third lived in Calcutta; many of the 183 non-Indian members (who constituted 62 per cent of the total membership) were resident in Europe or on furlough. Given that more than a third of the 112 Indian members also lived outside Calcutta, a high level of absenteeism was to be expected. Of the 145 Europeans for whom occupational information was given in the annual report, a quarter belonged to the ICS, an indication of the service's professional and recreational interest in the sciences, languages and arts of India and of the Society's own importance as a link between science and state. (The Society itself sometimes acted as a scientific pressure group, as in 1871 when its Council successfully petitioned the Government of India, in the wake of the *Challenger* expedition, to undertake deep-water marine research.) Among the remaining Europeans there were twenty members of the army medical services (principally the IMS) and an equal number of army officers, reflecting the continuing importance of the medical services and army to Indian science and engineering. Of the other Europeans, twenty-nine belonged to the scientific and technical services, along with smaller numbers of academics, lawyers, civil engineers, businessmen, doctors, churchmen and missionaries. Some, given their addresses, were likely to have been planters in Bengal and Assam.

The Indian membership was markedly different. Just over half were princes, zamindars and other notables; some, such as the Raja of Vizianagram, were significant patrons of science, but others clearly had a more decorative role. Only three Indian members belonged to the ICS (bringing the Society's total ICS membership to forty-two, or 14 per cent); but there were also five Indians in the scientific and technical services (bringing their tally to thirty-four or 11.5 per cent), as well as deputy magistrates, teachers, doctors and businessmen. Of the forty-six Indian members not identified by their profession, a large number were probably lawyers, minor landholders and government servants.[63]

If these figures can be taken as an approximate indication of the nature of the scientific community in India in the mid-1890s, then the dominance of the state (represented by the army, the ICS, the IMS and the scientific and technical services) is clear. But among the Indian members, the state was less in evi-

---

[62] *Proceedings of the Asiatic Society of Bengal, January to December, 1895* (Calcutta, 1896), pp. 50–51, 79, 91, 191, 200.   [63] Ibid., pp. iii–xiii.

dence; instead India's princes and landed elites held pride of place, though with what seems likely to have been a strong middle-class element. Overall, it is hard to escape the impression that in the capital of modern science in India the scope for Indian scientists was still remarkably small.

It might, however, be argued that the Asiatic Society was unrepresentative of the burgeoning interest in science across India as a whole. In late nineteenth-century India the engagement with science was certainly not confined to a few foreign-trained scientists and government servants, but was spread far more widely among India's old intelligentsia and its newer, Western-educated elites. The extent of this interest can partly be gauged through the existence of local societies and vernacular publications. One early example was Aligarh's Scientific Society, dating from 1864 and presided over by the Muslim 'moderniser' Sir Sayyid Ahmad Khan; another was the Bihar Scientific Society, established by Syed Imdad Ali, a sub-judge at Muzaffarpur, in 1868. The purpose of these societies was to spread the perceived benefits of modern science and technology through lectures and demonstrations, the translation of scientific and technical works and the publication of newspapers and tracts.[64] The Aligarh Society sought to translate scientific and technical works from English into the vernaculars, to locate and publish 'rare and valuable Oriental works', to publish any work 'which may be calculated to improve the native mind' and to present lectures on scientific and other 'useful subjects'. It translated some forty works on history, politics, trigonometry, algebra, arithmetic and especially agriculture, and in 1866 launched the *Akhbar Institute Gazette* to promote Western arts and sciences. Ahmad Khan planned to write an Urdu treatise on scientific methods of cultivation, but the project was never completed and the Society, from having had nearly 100 members in 1875, seems to have folded a few years later.[65]

Other societies and journals were published, particularly in Urdu and Bengali, to propagate Western medicine and sanitation or to integrate Western ideas with indigenous practices. The bi-monthly Urdu journal, *Guardian of Health*, published in Lahore from 1878 to 1888, sought to illustrate the principles of hygiene according to both 'the English and Native way of treating disease'.[66] Although such societies and journals were often short-lived and their effect localised, they were indicative of a widespread interest in bringing Western science and technology to a larger audience and in spreading its

[64] Kumar, *Science*, pp. 196–8.
[65] Hafeez Malik, *Sir Sayyid Ahmad Khan and Muslim Modernization in India and Pakistan* (New York, 1980), pp. 85–9.
[66] A. Neelameghan, *Development of Medical Societies and Medical Periodicals in India, 1780–1920* (Calcutta, 1963), pp. 75–6.

practical benefits and improving ideology. Although activities of this kind were to be found in most parts of India, Bengal, as the region where British ideas had their earliest and greatest impact, was particularly precocious and productive. As early as 1828 a Society for Translating European Sciences was established, and a Vernacular Literary Society followed in 1851. For several years Rajendralal Mitra ran the journal *Vividhartha Sangraha*, which sought to popularise geology, geography, physics and other sciences. Between 1868 and 1900, ten scientific and forty-seven technical journals were published in Bengal. Between 1875 and 1896, 776 science publications appeared in Indian languages in Bengal (more than half on medical topics), compared with 1,348 in all for Madras, Bombay, Punjab and the North-Western Provinces.[67]

The most momentous step in the creation of an Indian scientific community is often seen to be the founding of the Indian Association for the Cultivation of Science by Mahendralal Sircar in 1876. After receiving his M.D. in 1863, Mahendralal practised allopathic medicine for four years, but then rejected this in favour of homoeopathy, which he saw as a more rational medical system. Through the *Calcutta Journal of Medicine*, which he founded in 1868, Sircar called for greater Indian involvement in science, regarding it as the ultimate source of the 'superiority of the Western nations' and the key to India's progress and self-esteem.[68] In 1869 he published an article on the 'cultivation of science' in which he urged Indians to shake off their 'inherent indolence and apathy', become more self-reliant and look less to the state for their advancement. He commended the Asiatic Society (of which he was a member) for promoting science in India, but felt it had not done enough to help 'humble learners' develop an interest in science. A different kind of institution was required, one where public lectures and demonstrations would provide instruction for the masses and which was 'entirely under native management and control'.[69]

It took Sircar seven years to raise sufficient funds to launch the Indian Association for the Cultivation of Science. Public support for science was still clearly limited and the Association initially made little impact outside Calcutta. Ultimately, however, it was instrumental in establishing the principle that Indians were fit for 'pure' science and scientific research, and were not simply skilled artisans or departmental drudges without originality or initiative of their own. Where colonial science had been dominated by medicine and

---

[67] V. V. Krishna, 'The Colonial "Model" and the Emergence of National Science in India, 1876–1920', in P. Petitjean et al. (eds.), *Science and Empires* (Dordrecht, 1992), p. 64.

[68] *Indian Scientists: Biographical Sketches* (Madras, 1929), p. 34.

[69] Mahendralal Sircar, 'On the Desirability of Cultivation of the Sciences by the Natives of India', *Calcutta Journal of Medicine*, 2 (1869), pp. 286–91.

natural sciences, which focused attention on the investigation of the Indian environment, and even placed Indians themselves under scrutiny, Sircar directed Indian science towards physics, chemistry and other sciences that had little relevance to the externally perceived specificity of India and its peoples, but that helped Indians to connect with their own past and cultural legacies rather than to feel estranged from them. Although he saw science as having a practical role in Indian nation-building, Sircar was not content with the kind of functionalist approach that had dominated colonial science hitherto. Science had a 'higher and nobler claim than the narrow, utilitarian, Benthamite one. . . . It was the most powerful lever for progress, for the advancement of civilisation, for ennobling the mind of man'.[70] By the time of his death in 1904, Sircar had grown somewhat disillusioned with his grand project, lamenting the 'apathy of our people towards the cultivation of science'. But, the Association succeeded in taking science out of the narrow confines of the Asiatic Society and into a more public arena, in the process inspiring a new generation of scientists, including Jagadis Bose, Ray and Raman, and making Calcutta the undisputed centre of national science.

Other developments stimulated the growth of the Indian scientific community in these years. One was the rapidly changing political situation, especially the rise of militant nationalism and, in response to the Partition of Bengal in 1905, the Swadeshi movement. By arguing that India should produce its own goods rather than rely on British imports, proponents of the Swadeshi movement sought to strengthen India's drive for industrial self-reliance and, rather in contrast with Mahendralal Sircar's 'ennobling' view of science, emphasised the practical importance to India of science and technology. There had been moves in this direction even before 1905. In 1886 P. N. Bose published an influential pamphlet urging that more attention be given to technical and scientific education in Bengal, and five years later he established an Indian Industrial Association to promote the use of indigenous raw materials, though to little effect. The Association for the Advancement of Scientific and Industrial Education, founded by Jogendranath Ghosh in 1904, further sought to improve Indians' technical skills by financing their training abroad. In the same year the Dawn Society was launched to further the cause of national education. It took its name from the *Dawn* newspaper, founded in 1897 and an important vehicle for debates about the place of Western science and technology in India.[71] Under the stimulus of the Partition of Bengal, a series of new nationalist educational and technological initiatives arose. The Dawn Society

---

[70] *Indian Association for the Cultivation of Science: A Century* (Calcutta, 1976), p. 10.
[71] Dhruv Raina and S. Irfan Habib, 'The Unfolding of an Engagement: *The Dawn* on Science, Technical Education and Industrialization: India, 1896–1912', *SH*, 9 (1993), pp. 87–117.

became a National Council of Education, and Tarakanath Palit and Nilratan Sarkar launched the Society for the Promotion of Technical Education to implement an earlier project of P. N. Bose's for a Bengal Technical Institute. A rival faction within the National Council of Education formed the Bengal National College to further the cause of advanced scientific and literary studies. From these various initiatives eventually emerged the College of Engineering and Technology at Jadavpur and the University College of Science of Calcutta University.

Before 1900 little scientific research was conducted in Indian universities. Professorial chairs in botany, geology and chemistry were generally reserved for members of the IMS or the Educational Service. The great change came in the early twentieth century with the appointment by Curzon of a Universities Commission and the subsequent passing of the Indian Universities Act. This opened the way for universities to become centres of postgraduate training and research, able to appoint their own academic staff, maintain laboratories and museums and engage in all activities 'which tend to the promotion of study and research'.[72] In 1909, under its dynamic Vice-Chancellor, the lawyer and mathematician Asutosh Mukherjee, the University of Calcutta initiated postgraduate training and research in the sciences, creating over the next decade chairs in physics, chemistry, higher mathematics, botany and zoology.[73]

A further contributing factor was the role of Indian philanthropy, which helped to fund the rapid expansion of science in Indian universities and institutes and compensate for the colonial regime's relative indifference. When the government declined to finance the ambitious expansion of science teaching and research projected by Asutosh Mukherjee, Tarakanath Palit and Rash Bihari Ghosh donated Rs 2.4 million for the creation of chairs in physics and chemistry. Other wealthy Indians – landholders, lawyers and industrialists – provided scholarships and endowments that reduced financial and psychological dependence on the colonial state. Alongside British India's landed, commercial and industrial elite, the rulers or *diwans* (chief ministers) of princely states, notably Baroda, Hyderabad, Mysore and Travancore, showed a lively interest, in part prompted by considerations of political legitimacy and financial gain, in encouraging innovative scientific projects and in promoting technical education, as in the case of Baroda's Kala Bhavan Technical Institute.[74]

---

[72] Subbarayappa, 'Science', p. 566.

[73] P. C. Bagchi (ed.), *The Second City of the Empire: Twenty-Fifth Session of the Indian Science Congress Association, Calcutta, 1938* (Calcutta, 1938), p. 122.

[74] Dhruv Raina and S. Irfan Habib, 'Technical Content and Social Context: Locating Technical Institutes. The First Two Decades in the History of the Kala Bhavan, Baroda (1890–1910)', in Petitjean et al., *Science*, pp. 121–36.

One of the most conspicuous examples of Indian philanthropy in these years (and of the limitations to which it was subject) was the 3 million rupees' worth of property set aside by the Parsi industrialist Jamsetjee Nusserwanjee Tata for an Indian university of science in 1896. This was to be the Indian equivalent of the Imperial Institute in London, conducting research across a broad range of scientific fields but with direct relevance to the needs of Indian industry. Curzon, however, did not regard a science university as either practicable or desirable for India and steered the project towards narrower objectives. Bangalore was chosen as the site for the Indian Institute of Science, despite its remoteness from existing industries, partly because the Mysore state was prepared to contribute to its costs, but also because the climate was deemed suitable for the European staff who were expected to run it.[75] Eventually opened in 1908, the Institute, with departments for electrical technology, biochemistry and inorganic chemistry, was at first far from successful. It had difficulty in attracting students and in becoming a genuinely all-India institution; it was widely criticised by nationalists for not doing enough to further Indian needs in technical education and applied science, and even the colonial authorities found it poorly organised and without clear institutional aims. It stumbled on under European management until 1933, when C. V. Raman became its first Indian director and tried to re-establish it as a centre for research in his own field, physics.

The ambiguities surrounding the enterprise of science in India between the 1890s and 1920s are further illustrated by the founding of the Indian Science Congress in 1914. In some respects the Congress seemed to look backwards rather than forwards. The inaugural meeting was held at the Asiatic Society's premises and under its auspices. Noting in his presidential address that it was exactly 130 years since the founding of the Asiatic Society of Bengal, Asutosh Mukherjee paid tribute to Sir William Jones and to the Society, which, throughout its long career, had been 'the principal source of inspiration in the organisation and advancement of scientific research of every description in this country'. But, looking to the future, Mukherjee emphasised the importance of making India's landed classes 'realise that science enables us to solve difficult agricultural problems and thereby to revolutionise agricultural methods' and he appealed to the commercial community to recognise 'the inestimable value of science as an essential factor of industrial regeneration'. It was of great value for scientists to meet together to discuss their work and exchange ideas, he said, but it was no less important for them to 'bring their aims and views prominently into public notice' and 'press them upon the attention of the

[75] Home (Education), 5–9, February 1902, NAI.

government'. Even the 'most enlightened governments' occasionally needed to be reminded of 'the full extent of the paramount claims of science upon the public funds'.[76]

The initiative behind the founding of the Science Congress came not, as might have been expected, from Indian scientists in Calcutta, but from two Europeans recently appointed to chemistry chairs – Professor P. S. MacMahon of Canning College, Lucknow, and Professor J. L. Simonsen of Presidency College, Madras. Disappointed at the apparent dearth of research in Indian universities and the manner in which research seemed to be the prerogative of the state services, they sought (as had M'Clelland seventy years earlier) to create an all-India science forum like the British Association for the Advancement of Science. MacMahon and Simonsen appear to have been unaware how much scientific work was actually going on in India at the time, especially in Calcutta, but, to their credit, the Congress was a conscious departure from service-dominated imperial science. Divided into six sections – chemistry, physics, geology, botany, zoology and ethnology – the Congress gave a firmer institutional basis for science than the Asiatic Society had done. A third of the thirty-five papers presented in 1914 were given by Indians (four out of seven in physics, but only one of nine in zoology). Some papers echoed the old natural history of the Asiatic Society; but others signalled the emerging force of Indian research in physics and chemistry.[77] At first some of India's leading scientists seemed wary of the Congress, perhaps suspecting that it would prove to be yet another organisation in which Europeans lorded it over Indians. Of the first six presidents, only Asutosh Mukherjee was an Indian. But when P. C. Ray became its President in 1920, and delivered a staunchly nationalist address, and when the following year Raman took over as secretary, the Indian Science Congress was well on the way to becoming a vehicle for Indian scientists and representative of the expanding range of research in Indian universities. In time the Congress became as important a forum for national science as the Asiatic Society had previously been for colonial science.

CHEMISTRY, PHYSICS AND PLANT PHYSIOLOGY

Apart from mathematics, to which Asutosh Mukherjee and Srinivasa Ramanujan made important theoretical contributions between the 1880s and early 1920s,[78] the sciences that most clearly heralded the rise of the Indian

---

[76] *JASB*, 10 (1914), pp. lxxxix–xci.   [77] Ibid., pp. xcii–xciii.
[78] B. M. Sen, 'Progress of Mathematical Research in India', B. Prashad (ed.), *The Progress of Science in India during the Past Twenty-Five Years* (Calcutta, 1938), pp. 18–20.

scientific community were chemistry and physics. Chemistry had not been prominent among the sciences of the Company period. In 1883 when Pramatha Nath Bose compiled a survey of scientific papers presented to the Asiatic Society of Bengal since its inception, he found abundant material on geology and biology but hardly anything on chemistry. Until recently, he noted, there had been few laboratories in India worth the name, 'and we have had but few competent men with leisure to devote to lengthened chemical research'.[79] W. B. O'Shaughnessy, holder of the first chemistry chair at Calcutta Medical College, is better remembered as the pioneer of Indian telegraphy than for any original contribution to chemistry. His career typified the way in which the few chairs that existed in the physical sciences were generally held by the IMS rather than by individuals with a more specialised training. Except as an adjunct to medicine and geology, chemistry appeared to have little practical relevance to nineteenth-century India (even, as we have seen, to its agriculture), and facilities for its systematic study were almost entirely absent.

Towards the end of the century there were signs of change, with much of the initiative coming from Indian scientists. The very lack of British eminence and institutional control in such fields as chemistry and physics, and the absence of any dedicated state services, made them the more attractive and accessible to Indian researchers, in contrast to medicine where European dominance was vigorously upheld. Perhaps, too, as in the case of Mahendralal Sircar, the more theoretical nature of research in these sciences appealed to Indians' own intellectual legacies and aspirations more than botany and geology, the utilitarian field sciences that had grown up as part of the colonial tradition of exploration and exploitation. P. C. Ray and J. C. Bose were among the younger Indian scientists who pressed for the creation of new science courses and research facilities in Calcutta, just as it was Indian philanthropy, not state funding, that led to the establishment of the Palit chairs in chemistry and physics (reserved for Indians) and the creation of the University College of Science in 1916.

And yet by 1900 it was increasingly difficult for the British to ignore chemistry and its importance to the Indian economy. The development by German chemists of a synthetic dye to replace indigo was a severe blow to a once valuable export commodity, one that hit European planters particularly badly. For fear of adverse Indian comment, some effort was made to disguise the fact, but the establishment of an agricultural institute at Pusa, in the heart of the indigo country and with a former indigo expert as its first director, was

---

[79] P. N. Bose, 'Natural Sciences', in *Centenary Review of the Asiatic Society of Bengal from 1784 to 1883* (Calcutta, 1885), Part III, p. 101.

indicative of government's alarm at what Curzon dubbed the 'blue terror' of synthetic indigo.[80] In the wake of Voelcker's report, chemistry was belatedly recognised as having a direct contribution to make to the improvement of Indian agriculture. Chemistry was becoming vital to industry, too. The Indian Industrial Commission under Sir Thomas Holland expressed disquiet in 1918 at the neglect of chemistry in industrial research and development in India. It deplored the way chemists were currently employed in 'isolated posts', with different provincial and central government departments, and without any coordination of their research or an established career structure. It accordingly proposed the creation of a new all-India service for chemistry, modelled after Holland's old department, the Geological Survey, and recommended closer links between chemists and other branches of the scientific and technical services relating to agriculture and industry.[81]

When, however, a committee was appointed in 1920 to consider setting up an Indian Chemical Service it encountered fierce Indian opposition. P. C. Ray had long been critical of the scientific and technical services, with their privileged treatment of Europeans and close ties to the colonial regime. In 1899 he had deplored the way in which a European IMS officer might wake up one morning to find himself a Professor of Physiology 'through no qualifications of his own but solely by virtue of his seniority in the graded service'.[82] As a member of the 1920 committee, Ray did not accept the assertion, central to imperial science, that European recruitment was essential for the services to attract the best scientific talent; on the contrary, he regarded many European recruits as inferior to the products of Indian universities: India was already producing chemistry graduates who could 'hold their own against their confreres in any country in Europe'. He scorned proposals for an Indian Chemical Service, arguing that, whereas the creation of the IMS and ICS might have been justified a century or so earlier, the services had now become an 'anomaly, nay, a glaring anachronism'. A new all-India service would not serve the needs of Indian industry, but, on the contrary, would hamper the progress of science in India and stifle the productive and industrially useful research that had flourished in its universities over the previous quarter-century. It was these institutions that needed financial support, not new state services.[83]

An Indian Chemical Service did not come into being but, from his personal experience of the competing but deeply entangled worlds of science in the services and the universities, Ray was well placed to observe the growing, often acrimonious, rift between the two. Denied, despite his Edinburgh D.Sc., a post

[80] Kumar, *Science*, pp. 156–7.   [81] *Industrial Commission*, pp. 86–7, 92–3.   [82] Ray, *Essays*, p. 15.
[83] *Chemical Services Committee, 1920: Report* (Simla, 1920), pp. 105–15.

in the Educational Service, Ray was appointed instead to the provincial educational service, as an Assistant Professor of Chemistry at Presidency College, Calcutta, on a mere Rs 250 a month. He made his international reputation by his discovery of mercurous nitrite in 1895, his article in the *Journal of the Asiatic Society of Bengal* being reproduced in *Nature* in London the following year. In 1899, before the Swadeshi movement had formally come into being, he put nationalist principle into practice by establishing the Bengal Chemical and Pharmaceutical Works. From small beginnings, it became in time a highly successful business venture, particularly when the First World War cut off the supply of imported drugs and chemicals. By 1926 it employed 1,400 workers and had annual sales worth Rs 25 lakhs.[84] The company gave practical demonstration of one of Ray's other concerns – to show that Bengalis could run a large business and had the ability to combine laboratory research with industrial development. One can see why Ray did not believe in the need for state scientific services.

He also found a way to rival the authority of the colonial services by representing science as a patriotic duty and nurturing a nationalist *esprit de corps* among young Indian scientists. At Presidency College, and later at the University College of Science, Ray built up his own 'School of Indian Chemistry'. By 1920 he and his followers had published over 100 research papers, many of them in leading British and American journals, extending his work on mercurous nitrite into the study of other nitrites and metallic compounds. For all his Swadeshi convictions, Ray encouraged his ablest students to follow his own example and study abroad in order to acquire the best scientific training before returning to India. Many of his protégés went on to occupy important posts in chemistry and related departments in universities throughout India, creating academic networks that were in their way as extensive and influential as those of the state services. Ray and Jagadis Bose, both of whom in later life acquired the saintly image of *acharyas*, became scientific gurus. Their leadership had 'a quality reminiscent of that in the legendary *ashram* system of ancient India', with an 'intense personal relationship and loyalty between the leader as *guru* and the junior scientific workers in the group'.[85]

Ray turned chemistry into something of a subversive science. It was the kind of 'pure' science Indians were supposed to be unfit for, the kind of research for which they allegedly lacked discipline and dedication. It was, moreover, a

[84] J. Sen Gupta, *P. C. Ray* (New Delhi, 1972), p. 79.
[85] J. Mahanty, 'Science in the Universities since 1947', in B. R. Nanda (ed.), *Science and Technology in India* (New Delhi, 1977), p. 112.

science the British did not monopolise. Ray drew inspiration (as will be seen in the next chapter) from the metallurgy and alchemy of ancient and medieval India, but also from French chemists such as Marcellin Berthelot, from the Germans who, by pioneering synthetic dyestuffs, had scuppered the white planters of Bihar, and from the Japanese who had shown how an Asian society could adopt Western science and make it its own. Above all, Ray aligned science with nationhood. 'Our age is pre-eminently an age of science', he told the Indian Science Congress in his 1920 address. 'The fate of a nation will henceforth depend more upon the achievements of its students of science than upon the skill of its generals or the adroitness of its diplomatists and statesmen.'[86]

Physics did not lag far behind chemistry, and here Jagadis Bose played a role in some ways similar to Ray's, but in others significantly different. Returning from Britain in the mid-1880s, Bose was reluctantly admitted into the Educational Service, there being 'a strong doubt, not to say prejudice, against the capacity of an Indian to take any important position in science'.[87] Bose was appointed to a chair in physics at Presidency College and, in his own time, began research on electricity, extending Hertz's work on the properties of electromagnetic waves, experimenting with shorter wave lengths than had hitherto been investigated and devising new equipment to generate them. His work continues to be seen as remarkably innovative, though, disappointingly, the practical application of his work was left to others, including Marconi.[88] In 1896, following the publication of his paper on the polarisation of electric waves, Bose visited Europe on a 'scientific deputation'; he presented his findings to the Royal Institution in London and won copious praise. He was, Patrick Geddes observed, 'the first Indian to win distinction through investigation in science – in the most strictly Western of all its departments, and at that time also the most progressive'.[89] Bose's reception in London (matched by his appearances in Paris and Berlin) was remarkable; to the emerging scientific community in India it was inspirational. His original and remarkably sensitive work on electric waves demonstrated how much could be achieved with scant official support, with few laboratory facilities and without expensive equipment. In a sense, Bose (and to a lesser extent Ray) in the 1890s invented national science for India as laboratory science in contrast to the observational and field sciences that had been dominated by Europeans in India, and Bose in particular had the ability to make his dramatic experiments

---

[86] Ray, *Address*, pp. 1, 7–8.   [87] Geddes, *Pioneer*, p. 33.
[88] Visvapriya Mukherji, 'Some Historical Aspects of Jagadis Chandra Bose's Microwave Research during 1895–1900', *IJHS*, 14 (1979), pp. 87–104.   [89] Geddes, *Pioneer*, p. 62.

with electricity, as later with plants, appeal directly to Indian as well as European audiences, thereby overcoming much of the remoteness and stuffiness that had hitherto characterised Western science in India. Bose's success in London in 1896 led his sponsor, Lord Kelvin, and the President of the Royal Society, Lord Lister, to demand that Bose be given a laboratory in Calcutta worthy of his talents, but the Bengal government was unmoved. 'We often hear', Bose later remarked, 'that without a properly equipped laboratory, higher research in this country is an absolute impossibility.' He did not accept that view, but he well knew the costs involved. 'We often forget that the real laboratory is one's own mind,' he added. 'The room and the instruments only externalise that.'[90]

Around 1900 Bose moved into the second phase of his research career. Perceiving that the metal components in his electrical equipment seemed to suffer 'fatigue' and 'stimulation' like animate beings, Bose turned from physics to an investigation of the similar responses in the 'living' and 'non-living'. He developed a range of extraordinarily sensitive laboratory instruments, such as a 'resonant recorder' capable of recording responses at intervals of one-thousandth of a second. By detecting 'plant signatures' and making plants 'speak', Bose showed that plants (and even metals) were responsive to heat, light, pain and disturbance in much the same way as humans and animals. But when Bose presented his new work on biophysics in Europe, he met with scepticism, even scorn, from British physiologists, who derided the suggestion that there could be any meaningful correspondence between human 'responses' and plant 'reactions'. Bose was suspected of having succumbed to his native mysticism and the fanciful flights of an 'Oriental imagination'. This rejection blighted Bose's international career and possibly cost him a Nobel Prize.[91] Nor was this an exclusively Western reaction: C. V. Raman also reputedly called Bose's work on plant physiology 'mumbo-jumbo'.[92]

In some respects, it was Raman who was the principal beneficiary of Bose's pioneering labour. Bose had begun the task of winning recognition for Indians in science both at home and abroad, and Raman was able to capitalise on this. Raman's own work on physics relied, as Bose's had done, on fairly basic laboratory equipment – this was still the 'sealing wax and string' era of physics research and it was possible for men like Bose and Raman to conduct significant research without expensive resources, large research teams or the

---

[90] *Sir Jagadis Chunder Bose: His Life and Speeches* (Madras, 1920), p. 90.
[91] Deepak Kumar, 'The "Culture" of Science and Colonial Culture, India, 1820–1920', *BJHS*, 29 (1996), pp. 204–5.
[92] Krishna Dutt and Andrew Robinson, *Rabindranath Tagore: The Myriad-Minded Man* (New York, 1996), p. 129.

support of an advanced industrial base.[93] In 1930 Raman won the Nobel Prize for Physics for his work on the diffusion of light in liquids, dubbed the 'Raman effect' by analogy with the 'Compton Effect', which had won the Physics Prize for its American inventor three years earlier. Bose, by contrast, never won the ultimate accolade of international science and in his later years was often regarded as having failed to fulfil the immense potential of his early work on electric waves. Many questions remained unanswered, too, about his 'bold hypothesis' about plant responses.[94] Nevertheless, Bose remained an inspirational figure to the wider public, especially for his eloquent Bengali writings, and among the Indian scientific community at large. A 'poet among biologists', Bose was a scientific Romantic, who, like Humphrey Davy at the Royal Institution several decades before Bose went to lecture there, sought to use the power of intuition and imagination to uncover a meaning and interconnectedness in the hidden properties of nature, and who saw the need to popularise science in India and to give it a new, more culturally integrated, role.[95] He was long remembered as having brought Indian science, after centuries of absence, back to the centre of the world stage. The botanist Birbal Sahni remarked on his death in 1937 that 'an incredibly long period of degradation separated us from a great and proud past. Indians were known only as dreamers and philosophers, their right to be heard as scientists only laughed at'. Bose was 'the pioneer who broke this spell'.[96]

[93] John Ziman, *The Force of Knowledge: The Scientific Dimensions of Society* (London, 1976), pp. 272–3.
[94] J. C. Ghosh, 'Sir Jagadish Chandra Bose', in Rushbrook-Williams (ed.), *Great Men*, pp. 588–9.
[95] Ghosh, 'Bose', p. 589; cf. David Knight, 'Romanticism and the Sciences', in Andrew Cunningham and Nicholas Jardine (eds.), *Romanticism and the Sciences* (Cambridge, 1990), pp. 13–24.
[96] M. N. Saha, 'The Problem of Indian Rivers', *Annual Address to the National Institute of Science* (Calcutta, 1938), p. 3.

CHAPTER 6

SCIENCE, STATE AND NATION

In Stockholm on 10 December 1930, C. V. Raman received the Nobel Prize for Physics for 'the discovery of the effect named after him'. The prize had previously been awarded to such renowned physicists as Marie Curie, Albert Einstein and Niels Bohr, but never before to a non-European. Ronald Ross won the Medicine Prize in 1902, and Rabindranath Tagore, the only other Indian to receive a Nobel Prize before 1947, the Literature Prize in 1913, but it had long been seen as a matter for regret, as one Indian observer put it in 1912, 'that none of our country's scientists have up till now been awarded this much coveted prize'. He hoped that in the near future at least one of his compatriots would win 'this blue ribbon in science' and so achieve 'the regard of the world'.[1] That winter's afternoon in Stockholm in 1930 it seemed Indian science had finally won that 'regard'.

Along with the acclaim for Bose's work on electric waves in London in 1896 and the founding of the Indian Science Congress in 1914, Raman's Nobel Prize in 1930 could be seen as a symbolic milestone in the emergence of national science in India, the point at which it finally broke free from British tutelage and control. Certainly, by the 1920s and 1930s science in India had attained a new maturity and authority and Indian scientists had begun to acquire both international recognition and positions of intellectual and institutional leadership within India. And yet, in many respects, India's science remained constrained and conditioned by the continuing presence of colonial rule and troubled by uncertainties about status and identity in India's quest for nationhood and modernity.

THE RETURN OF HINDU SCIENCE

It is a striking paradox that, at the very time when Western science and medicine in India seemed finally to have established an unassailable superiority over indigenous rivals, they were confronted by a powerful revival of indigenous, and more especially 'Hindu', science. This resurgence can be understood in two ways. It can be seen as a direct nationalist riposte to the alien, coercive,

[1] M. Singaravelu, *The Interpretation of Science* (Madras, 1914), p. 17.

state-centred nature of Western science and medicine. There was certainly an element of this, but the revival of Hindu science was often less a direct challenge to Western scientific hegemony than a complementary cultural and political activity, a means of establishing Indian antecedents and contexts for modern science and securing the self-esteem and autonomy of Indians in the contemporary scientific world. It was thus no accident that the revival of Hindu science coincided with a new, more confident phase in the history of Western science in India and the emergence of an Indian scientific community; these were, to a striking degree, interdependent phenomena.

The authority that Western science had come to enjoy in India by the late nineteenth century was too great to be ignored in Indians' own programmes of reform and revitalisation. Wilhelm Halbfass has argued that, for the Hindu tradition in particular, 'the encounter with modern science and technology was the most momentous part of its encounter with the West', though he recognises that there was no single response.[2] There were at least two ways in which Indians sought to come to terms with Western science. One was to see it as partly fulfilling the promise of India's own antiquity while, at the same time, representing a Western materialism that was ultimately inferior to the spiritual and intellectual legacy of the Indian religious and philosophical tradition. David Gosling has suggested that there was little opposition to the Darwinian theory of evolution in India because reformers like the Brahmo leader Keshab Chandra Sen ignored the struggle for survival that lay at its heart, seeing evolution instead as an affirmation of the idea of a progression, represented by the avatars of Vishnu, from primordial to higher states of being. An even more extreme and influential attempt to appropriate the West to the East was made by Dayananda Saraswati, founder of the Arya Samaj in 1875, who claimed that the discoveries of modern science had been anticipated in the Vedas as the fountainhead of all science and religion.[3] Vivekananda, too, saw 'a good deal of similarity between Indian thought and modern science'. 'The moderns have their evolution', he told an American audience in 1896, 'and so have the Yogis.' They, too, had seen how one species evolved into another. But the evolution of the Yogis was superior because it was primarily concerned with spiritual evolution and not merely changes in outward form. Indeed, the 'horrible idea of competition', rather than being an essential evolutionary dynamic, as Darwin had imagined, stood in the way of realising higher, spiritual goals.[4] A writer in *Dawn* took a similar view in 1897 when he described how science in

[2] Wilhelm Halbfass, *India and Europe: An Essay in Understanding* (Albany, NY, 1988), p. 399.
[3] David L. Gosling, *Science and Religion in India* (Madras, 1976), pp. 42–7.
[4] *The Complete Works of Swami Vivekananda* (9th edition, Calcutta, 1970), pp. 277–8.

the West had merely fed material wants and desires and accentuated the division of European society into virtual armed camps. Ancient India showed how man could be the master, not the slave, of science, but this called for 'the cultivation of the higher science of the spirit' over the 'lower science' of material objects.[5]

Setting up Western science and technology as antithetical or inferior to Eastern spirituality was only one of the ways in which Hindu writers addressed the problem of science. Another of the striking characteristics of Indian thought in the late nineteenth and early twentieth centuries, especially in Bengal, was the importance attached to history as a means of comprehending India's colonial predicament and of seeking ways to subvert colonial authority; science was no exception to this. There were, however, specific reasons why science should be preoccupied with history. One was the manner in which Western science presented itself historically, as heir to a long tradition of invention and discovery. Another, more immediate, influence was the example of Western Orientalists and the picture they presented of the early achievements and subsequent decline of Hindu civilisation. Indian writers were concerned to establish why Indian science had been part of this decline and sought ways to compare and connect the history of their science with the hegemonic Western tradition. Increasingly, this turned into an attempt to establish the distinctive character of Indian science and the unique contribution it had made, and therefore could again make, to the wider scientific domain.

The need to historicise science came not least from scientists themselves. In 1884 Pramatha Nath Bose set out to present a survey of scientific articles published in the journals of the Asiatic Society since its founding 100 years earlier but also to situate Hindu science relative to the Western tradition. Bose's account inevitably began with Jones and the early Orientalists, and here most scientific fields appeared to have no prior Indian history worthy of comment. The history of astronomy (oddly, in view of its eminence among the sciences of pre-colonial South Asia), trigonometry and meteorology were all presented as a history of science in Europe or of Europeans working in India. Only when he turned to mathematics did Bose introduce his other concern, beginning this section of his survey with the observation that the mathematical sciences 'had been cultivated by the Hindus from very remote times'. However, 'the history of the progress and civilisation of that nation' had ended in the twelfth century, Bhaskaracharya being 'the last of the noble band of mathematicians which it is the glory of this country to have produced'. Thereafter 'the degenerated

---

[5] 'Side-lights on Modern Science', *Dawn* (June 1897), pp. 99–100.

Hindus' wasted their time in 'inventing an infinity of absurd, frivolous stories about gods and goddesses, demigods and *avatars*'. They 'forgot the principles of their sciences', which were thereby 'reduced to mere arts practised by ignorant astrologers and needy physicians'.[6] Bose found few exceptions to the 'general degeneracy of the Hindus since the thirteenth century'. And yet the story of Hindu degeneration was also ultimately a tale of redemption, for 'to rescue those sciences from oblivion was one of the noble objects which the Asiatic Society set itself from its very foundation to accomplish'.[7] This recovery and the insertion of Hindu science into what was essentially a narrative of colonial science were, however, only a temporary diversion. When, a few pages later, he moved on to geology and botany, Bose began again with the history of those sciences in the West. Most strikingly of all, when Bose turned to chemistry, he saw no opportunity to supplement or contrast the Society's meagre record with an account of the achievements of ancient Hindu chemistry.

This was a task left to Prafulla Chandra Ray. His *History of Hindu Chemistry*, published in two volumes in 1902 and 1908, firmly established in many Indian minds the scientific credentials of the ancient Hindus and, by extension, their rightful place in the modern world of science. Significantly, Ray was not a professional historian; writing history, he claimed, was only indulging his 'favourite hobby'. But it is hard not to see the *History* as a conscious attempt to build up self-esteem within the Indian scientific community and to gain national and international recognition for India's scientific tradition. The first volume appeared shortly before the launching of the Swadeshi movement and coincided with Ray's attempt to establish the Bengal and Chemical Pharmaceutical Works. The *History* could not have been more opportune in linking the inspiration and achievements of the past with the needs and aspirations of the present, and it marked a decisive break from the hesitancy and ambiguity of Bose's survey twenty years earlier. The focus on *Hindu* chemistry was no accident: a notion of 'Hindu' was central to a work that consistently argued for the originality and vitality of an indigenous scientific tradition that owed nothing to the West or Islam. Drawing upon largely unpublished works, Ray demonstrated the extent to which religion had imparted a distinctively Hindu character to Indian chemistry, which had first emerged in India as the 'handmaid' of Ayurveda and was later developed as an alchemical adjunct to medieval Tantric cults.[8]

---

[6] Pramatha Nath Bose, 'Natural Science', in *Centenary Review of the Asiatic Society of Bengal from 1784 to 1883* (Calcutta, 1885), pp. 20–1. [7] Ibid., pp. 21, 25.

[8] Prafulla Chandra Ray, *A History of Hindu Chemistry from the Earliest Times to the Middle of the Sixteenth Century AD*, 2 vols. (London, 1907, 1909; first published 1902 and 1908 respectively).

But, although Ray saw the history of Hindu science as a source of national pride and inspiration, he did not see it as something that could be revived as an alternative to modern science; the latter alone could secure India's modern nationhood. In his address to the Indian Science Congress in 1920, Ray stated that, despite its early scientific achievements, India had stagnated and lapsed in medieval times into 'silent and ecstatic mediation', while science in Europe had advanced through Galileo, Newton and Boyle. He did not blame the Muslims for this decline, which had begun before their arrival. Rather, Ray, in his Brahmo rationalism, believed that an 'intellectual torpor' had taken possession of 'the Indian mind'. Reverence for the *shastras* had taken 'the place of reason' and was 'inimical to the study of science, which accepts things not on trust, but by verification'.[9] The arrival of the British had helped end India's scientific dark age, but Indians could still learn from the scientific rigour of the West. As Ray remarked a few years earlier, 'It is to Europe that we must now turn our eyes for the realisation of the ideal presented by our own *rishis* – unflagging and concentrated devotion to the pursuit of knowledge for its own sake.'[10]

Ray's momentous claims for Hindu chemistry were taken up and developed by others, most notably Brajendranath Seal. Professor of Philosophy at Calcutta University, Seal contributed an essay on the 'mechanical, physical and chemical theories of the ancient Hindus' to the second volume of Ray's *History*, before publishing his own extended account of *The Positive Sciences of the Ancient Hindus*. Like Ray, Seal drew heavily on Ayurvedic texts to retrieve the methodological and technical achievements of ancient India. He showed how the Hindus, 'no less than the Greeks', had participated in the task of constructing scientific concepts and methods in the ancient world and how they had built up a vast body of empirical knowledge that, once applied to metalworking, textiles and other forms of manufacturing, had made India, for centuries, one of the world's great trading nations. Seal acknowledged that in some fields the early Hindus' scientific understanding was defective: it was clear, for instance, that Caraka and Susruta had 'no idea of the part played by the lungs in the purification of the blood'. But in general he sought to elevate the Hindus' contribution to science above mere empiricism by emphasising their 'rigorous scientific method'. The whole movement of Hindu science, he believed, was 'genuinely and positively scientific', though arrested at an early stage of its development.[11]

Seal, too, saw evidence for a theory of evolution in early Indian thought, and

---

[9] Prafulla Chandra Ray, *Indian Science Congress Presidential Address* (Calcutta, 1920), pp. 1–3.
[10] Prafulla Chandra Ray, *Essays and Discourses* (Madras, 1918), p. 204.
[11] Brajendranath Seal, *The Positive Sciences of the Ancient Hindus* (Calcutta, 1915), pp. 215, 244–8.

credited Bhaskaracharya and Vachaspati with having anticipated by centuries the work of Newton and Descartes. He also, like many of his contemporaries, saw a direct correspondence between philosophical terms in Sanskrit texts and recent scientific discoveries in the West. A science that spoke of atoms, molecules and particles offered fresh opportunities for those who sought to establish direct correspondences between the wisdom of ancient Hindu sages and the atomic theories of Dalton, Einstein and Bohr. In this endeavour Seal was not as rash as some of his contemporaries, and in 1915 warned that what the *rishis* had was 'felicitous intuition earned . . . by intense meditation and guided by intelligent observation', but 'the gulf between this stage and the positive experimental knowledge of science is profound, and cannot be traversed except by means of difficult and delicate methods of quantitative analysis and measurement', such as Jagadis Bose was currently undertaking through his studies of plant physiology.[12] Critics, none the less, saw in Seal, as in Ray, an 'unhistorical and uncritical' approach to obscure and undatable texts and a failure to distinguish between scientific observation and philosophical conjecture. Both displayed a 'somewhat obsessive anxiety to attribute as much as possible to the ingenuity of the Aryans', while providing little tangible evidence in support of their 'sweeping claims'.[13]

Ray, whose investigation of mercurous nitrite preceded his discovery of the important role of mercury in Tantric alchemy, saw Hindu science as a source of inspiration rather than practical guidance. But other Indians saw the philosophical and scientific legacy of the ancient Hindus as having a more immediate bearing on modern science. Jagadis Chandra Bose in particular sought an Indian modernity that was different from the West's even within the supposedly universal discourses and practices of modern science. He did this largely by drawing from Hinduism what he saw as a unique insight denied to the West (though one that had parallels with the scientific Romanticism of the early nineteenth century) – the underlying unity of all forms of existence. There was no 'absolute barrier', he told a London audience in 1901, between the physical and the physiological, and his work on plants demonstrated the validity of the message 'proclaimed by my ancestors on the banks of the Ganges thirty centuries ago – "They who see but one, in all the changing manifoldness of this universe, unto them belongs Eternal Truth – unto none else, unto none else!"'.[14] Bose saw no contradiction between Hindu philosophy and the methodological exactitude required in modern scientific research. Indeed,

---

[12] *MR* (March 1918), p. 328. [13] Gosling, *Science*, p. 25.
[14] Patrick Geddes, *An Indian Pioneer of Science: The Life and Work of Sir Jagadis C. Bose* (London, 1920), pp. 97–8; cf. Partha Chatterjee, 'Our Modernity', in *The Present History of West Bengal: Essays in Political Criticism* (Delhi, 1997), p. 208.

India was 'the only country where so far from there being a conflict between science and religion, knowledge is regarded as religion itself'.[15] With their keen imagination, intuition and understanding of the interconnectedness of all life forms, Indians were in many respects better suited to sensitive research than were Western scientists, whose approach was aggressive and crudely materialistic, and whose tendency constantly to subdivide scientific fields precluded them from seeing the underlying unity. Far from the West having a monopoly of scientific ideas and insights, India, Bose believed, had much of its own to contribute, now as in the past, and he expected the international scientific community to respect and appreciate this. Ancient seats of Indian learning, such as Taxila and Nalanda, served for him as evidence of the free and equal exchange of ideas that had once existed between different societies and cultures. He told an audience at Lahore in 1913 that 'although science was neither of the East nor of the West, but international in its universality, certain aspects of it gained richness of colour by reason of their place of origin'. India, 'through its habit of synthesis', was able 'to realise instinctively the idea of unity and to see in the phenomenal world an universe instead of a multiverse'.[16]

However critically he might be seen in the West, in India Bose inspired other scientists to try to establish connections between Western and Hindu science, and to use the latter to inform and enrich present-day science. There was a new interest in scrutinising Ayurvedic texts for what they might reveal of the scientific knowledge of the ancients and contribute to modern medicine, botany and zoology. One example of this was K. R. Kirtikar and B. D. Basu's *Indian Medicinal Plants*, published in 1918, which attempted a fresh synthesis of Western materia medica and Ayurvedic botany. The authors, both formerly of the IMS, acknowledged a debt to Roxburgh, Royle and Hooker, but identified most strongly with an Indian tradition of medical botany going back to Caraka and beyond. It was, they held, 'greatly to the credit of the people of India that they were acquainted with a far larger number of medicinal plants than the natives of any other country on the face of the earth'.[17] Similarly, in 1927 a Professor of Botany at Presidency College trawled Sanskrit texts for references that might reveal useful insights into the ancients' knowledge of plant life, and in 1935 a government zoologist delved into Susruta to show how advanced had been the ancient Hindus' understanding of the relationship between form and function in fishes.[18]

[15] *Sir Jagdis Chunder Bose: His Life and Speeches* (Madras, 1920), p. 149. [16] Ibid., p. 118.
[17] K. R. Kirtikar and B. D. Basu, *Indian Medicinal Plants*, 1 (Allahabad, 1918), pp. xxxiii–xxxv.
[18] Girija Prasanna Majumdar, *Vanaspati: Plants and Plant-Life as in Indian Treatises and Traditions* (Calcutta, 1927); Sunder Lal Hora, 'Ancient Hindu Conception of Correlation between Form and Locomotion of Fishes', *JASB (Science)*, 1 (1935), pp. 1–7.

In all these works, ancient science was seen to anticipate modern science or to offer still useful insights and information. Hindu knowledge and modern science were seen to exist in symbiosis. Nor was this appreciation of the Hindu contribution confined to the physical sciences and ancient texts. Girindra Shekar Bose, the founder of Indian psychology, argued in 1938 (rather along the lines of J. C. Bose and Seal a quarter century earlier) that ancient India's men of learning had 'had a genius for introspective meditation'. The modern Indian psychologist was heir to this tradition, which, far from being evidence of a proclivity for empty speculation, gave the Indian practitioner a distinct advantage over Western colleagues. If this faculty were 'properly cultivated', then analytical problems requiring 'deep introspection' (such as those concerning 'thought processes, higher cultural inhibitions, etc.') could be successfully solved. Bose believed that the 'mystical experience of saints and yogis' should itself be taken up as a subject for psychological research and that India was the best place for such a study.[19]

THE REVIVAL OF INDIGENOUS MEDICINE

The most practical, and therefore most contentious, area of engagement between Hindu and Western science lay in the field of medicine. At precisely the time when Western medicine was seeking a new authority through bacteriology, parasitology and interventionist public health programmes, there was a growing interest in India in the revival and revitalisation of indigenous medicine. Medicine and medical texts such as the Caraka *samhita* were central to establishing claims for the validity and rationality of Hindu science as a whole, and Ayurvedic medicine could be seen as having a practical value in the modern age and having insights and pharmacological knowledge that Western medicine had still to uncover. Its claims to be a 'system', recognised as early as the 1840s by T. A. Wise in his *Commentary*, implied an equality with the Western system and allowed equivalent claims for recognition to be made for the Unani, Siddha and other 'systems' of medicine, even for homoeopathy, despite its European origins. Moreover, although Ayurvedic texts were sometimes interpreted (as by Mahendralal Sircar in the 1860s) as essentially secular works, the revivalist movement, which began in earnest in the 1890s, was largely a corollary of the rise of Hindu nationalism. This identified Ayurveda with the ancient Aryans, even though in practice what it aspired to was the integration

---

[19] G. Bose, 'Psychology', in B. Prashad (ed.), *The Progress of Science in India during the Past Twenty-Five Years* (Calcutta, 1938), pp. 345–6.

of selected aspects of Western medicine with a reformed Hindu medicine rather than the pursuit of 'pure Ayurveda'.

An early and widely noted contribution to the revivalist movement was the publication in 1895 of a treatise on *Aryan Medical Science* by Bhagvat Sinhji, Maharaja of Gondal. After the manner of the Orientalists and the Arya Samaj, Sinhji took up the theme that the Aryans, 'the most enlightened race in the dawn of history', had been pioneers of several scientific fields, including medicine. Hindu medicine had reached 'the acme of its glory in the time of the Ramayana and the Mahabharata', Sinhji asserted, but declined with the coming of the Muslims and the support they gave to their own (Unani) system of medicine – a clear lesson that 'no art or science can flourish without the moral and material support of the government of the day'. Like many of those whom Charles Leslie has identified as syncretists and integrationists,[20] Sinhji did not expect revived Ayurveda entirely to supplant the Western system. He looked instead for mutual respect, believing that 'the Hindoo system of medicine can, on the whole, bear comparison with the Western'. He urged the 'Western and Eastern Schools of Medicine' to 'join hands and reconcile themselves to each other wherever possible'. They should meet 'as friends, and not as foes or rivals. ... the East has much to learn from the West, but the West, too, may have something to acquire from the East, if it so chooses'.[21] Twenty years later, Kaviraj Gananath Sen, 'perhaps the best-known advocate ... of an integrated medical system',[22] responded to the charge that Ayurveda was 'a very antiquated ... empirical and unscientific method of treatment'. Western critics would do well, he said, to learn Sanskrit before they tried to 'cast shame on Ayurveda' by misrepresenting its 'great truths', including its humoral pathology. Like Sinhji, Sen did not deny the need to learn from allopathy, but he did not see this as detracting from the valuable and distinctive nature of the Hindu system as a whole. 'The spirit of Ayurveda' was nothing less than 'the spirit of progress. It demands that you should interrogate and learn and add to the store of knowledge left to you by the Ancients'. India's salvation lay in 'the rejuvenation of Ayurvedic medicine side by side with the progress of the Western system'.[23]

It was a further reflection of the authority Western medical techniques had acquired that proponents of Ayurveda asked that its drugs and methods be recognised as fully scientific. It was argued that their value had been established by virtue of their having been tried and tested over many centuries, but

---

[20] Charles Leslie, 'Interpretations of Illness: Syncretism in Modern Ayurveda', in Charles Leslie and Allan Young (eds.), *Paths to Asian Medical Knowledge* (Berkeley, CA, 1992), p. 179.
[21] Bhagvat Sinh Jee, *Aryan Medical Science: A Short History* (Delhi, 1993), pp. 22, 187, 200, 205.
[22] Leslie, 'Interpretations', p. 183.   [23] Kaviraj Gananath Sen, *Hindu Medicine* (Madras, 1916).

suggested that modern laboratory techniques should be used to confirm their scientific value by modern methods. Although in several respects Hindu science and the Western view of tropical 'otherness' represented contrasting approaches, advocates of Ayurveda took up the idea of India as tropical to argue for the greater efficacy of their medical system. Since, it was said, Ayurvedic drugs were themselves the product of a tropical country and had been used for centuries on a 'tropical race', they were better suited for use by Indians than drugs developed in the cold climate and 'alien' cultural conditions of temperate Europe.[24]

The Ayurvedic revival assumed several forms. The location and translation of Ayurvedic texts from Sanskrit into English gave the movement textual authority and made its major works accessible to a wide audience.[25] Ayurveda established strong local roots, for instance by setting up dispensaries, thereby emulating one of Western medicine's most distinctive and effective institutional forms. The first Ayurvedic dispensary in Calcutta was opened in 1878 by Kaviraj Chandra Kishore Sen; the earliest in Madras was founded twenty years later by the Telugu *vaid*, Pandit D. Gopalacharlu. From the 1890s a number of Ayurvedic dispensaries, along with those for Unani medicine and homoeopathy, were subsidised or run by municipal councils. No less significant in popularising the new indigenous medicine was the use of the printing press, as newspaper advertisements and articles, tracts and medical journals fed the literate public with extracts from Sanskrit texts and plied them with information about indigenous remedies and patent medicines. The apparent inability of Western medical practitioners to deal promptly or appropriately with cholera, malaria, plague and influenza, or with more intimate diseases such as syphilis, encouraged recourse to Ayurvedic, Unani and homoeopathic remedies. No less significant, from the practitioner's viewpoint, was the apparently discriminatory attitude of the colonial medical establishment. Among those who took up Ayurveda in Andhra in the late nineteenth and early twentieth century were many practitioners who had trained in the Western system but resented the lack of status and income it afforded them. Rather than be confined indefinitely to the ill-paid, lower ranks of the state medical system, they preferred to strike out on their own, eclectically combining elements of allopathy with indigenous therapeutics. Given the popularity, accessibility and relative cheapness of the latter, this syncretic approach brought

[24] T. R. Ethirajulu Naidu, *The Ayurvedic System* (Madras, 1918), p. 10; G. Srinivasa Murti, 'A Memorandum on the Science and Art of Indian Medicine', Appendix I to *The Report of the Committee on the Indigenous Systems of Medicine, Madras* (Madras, 1923), pp. 95–6.

[25] Gurdip Singh and P. D. Joshi, 'Impact of European Science and Technology on the Development of Modern Ayurveda during 19th Century', *IJHS*, 17 (1982), pp. 313–25.

them more income, respect and influence than if they had stayed in state service.[26]

The indigenous revival was a war on two fronts, seeking to re-establish Ayurveda as the popular and culturally appropriate alternative to allopathy, but also seeking to supplant what were seen as ignorant and superstitious folk practices. An important illustration of this dual process is provided by the career and ideas of P. S. Varier, a leading figure in the Ayurvedic movement in Kerala. Born in 1869, Varier studied for four years under one of Kerala's most respected *vaids*. But by the time he was 20 Varier had decided his ambition would be to try to 'raise Ayurveda', which had fallen into disrepute, 'to higher levels and bring it in line with the popular system of allopathy'. He learned English and acquired a knowledge of Western medicine from a sympathetic allopathic doctor, who was an assistant surgeon at the government hospital in Manjeri. Varier established a practice at Kottakkal in Malabar, where he sought to demonstrate to patients the 'valued qualities and efficacy of the Ayurvedic medicines'. In 1902 he launched the Arya Vaidya Samajam, drawing up his own manifesto for an Ayurvedic revival. To regain popularity, he argued, Ayurveda had to 'move out of the old ruts, adopting modern techniques without detriment to its inherent qualities'. He blamed *vaids* for having come to rely on illiterate apothecaries to prepare their drugs, with the result that patients received ineffective medicines and lost confidence in the system as a whole. It was necessary, therefore, that *vaids* should learn to prepare medicines themselves and to maintain the standards of the drugs they used. Varier urged *vaids* to join together to manufacture their own medicines 'on strict Ayurvedic principles', and then supply them to other practitioners or directly to patients. This was another area where *vaids* could learn from allopaths, who received their medicines from stockists and pharmaceutical firms. 'It is meaningless', Varier explained, 'to say that our ancestors had not adopted such methods. It is time we moved with the times and followed modern ways, wherever useful and necessary'. Varier put his ideas into effect by setting up an Ayurvedic pharmaceutical company, which, by the time of his death in 1944, had monthly sales worth Rs 30,000.[27]

From such local beginnings and grassroots interactions with Western medicine, the Ayurvedic movement evolved into regional and all-India organisations. When the Arya Vaidya Samajam met in 1902 at Kottakkal under Varier's leadership it attracted the support of the maharajas of Travancore and Cochin

---

[26] Poornima Sardesai, 'Indigenous Medical Practitioners and Colonial Rule in India: A Study in the Production of Medical Knowledge in Andhra, 1880–1930', unpublished paper, 1996.

[27] Kizhedath Vasudevan Nair, *A Biography of Vaidyaratna P. S. Varier* (Kottakkal, 1954), pp. 20–35.

and the Zamorin of Calicut – the patronage of princes and landed magnates, as well as the professional middle classes, was important in giving the movement respectability and the authority to match the colonial state's support for allopathy. As K. N. Panikkar has shown, the attempt to reform and revitalise Hindu medicine was inseparable from broader tides of cultural and intellectual regeneration. Varier helped to promote the Malayali language and drama; his journal *Dhanwantari*, launched in 1902, became the 'mouthpiece of the revitalisation movement in Kerala'. Varier wrote a series of Malayali medical texts that, along with the medicines themselves, made Ayurveda more accessible to the people. At the same time, reformed Ayurveda was squeezing out 'unscientific' practitioners, replacing them with its own brand of syncretic medicine. While seeking to establish their position relative to the politically dominant allopathy, the proponents of Ayurveda were also pursuing their own quest for cultural hegemony.[28]

The movement for revival and reform was not confined to Ayurveda but extended to Unani medicine as well. Hakim Ajmal Khan, the son of a celebrated physician, took over the Madrassa Tibbia in Delhi in 1889 and sought to use it to revive and modernise the Unani system. He was aided by donations from Muslim notables such as the Nawabs of Rampur and Dacca, but also sought state recognition and support. In a petition to the government in 1915, Ajmal Khan emphasised that Western and Unani medicine were really one and the same, both having their origins in Greek medicine. That system had been brought to India by the Muslims and flourished under the Mughals, but when their rule declined so did the Unani system. Like the Ayurvedic modernisers, Ajmal Khan sought to collect and systematise traditional medical knowledge but also wanted 'to do away with the unqualified Tabibs [doctors] who infect the various parts of India to the great detriment of the health of the inhabitants'. In order to train properly qualified physicians, his Madrassa gave instruction in practical anatomy and surgery as well as Unani diagnostics and pharmacology. In addition, Ajmal Khan sought to open a dialogue with Ayurveda and to incorporate its teachings into his reform programme. By 1915 the Madrassa, now expanded into an Ayurvedic and Unani Tibbi College, had trained 263 students, the great majority of them Muslims.[29] But Ajmal Khan's attempt to modernise Unani did not please traditionalists in Lucknow, where a more conservative school of *hakims* existed and sought to organise support for 'pure' Unani medicine. Ajmal Khan's college, one critic lamented, included

---

[28] K. N. Panikkar, 'Indigenous Medicine and Cultural Hegemony: A Study of the Revitalization Movement in Keralam', *SH*, 8 (1992), pp. 283–308.
[29] Ajmal Khan to Sir Pardy Lukis, Home (Medical), 41, July 1916, NAI.

'so many European medical practices that the original science of Muslim medicine, instead of advancing, seems headed for disaster . . . the future of our ancient medical science in Delhi appears to be greatly endangered'.[30]

Like other reformers, Ajmal Khan was alarmed at the passing of provincial medical registration Acts (beginning with Bombay in 1912), which gave recognition to practitioners of Western medicine but excluded all others. The fear that all practitioners of indigenous medicine would be treated as 'quacks', regardless of their modern affinities and training, was seemingly reinforced in 1915–16 when an Indian allopathic doctor in Madras was removed from the medical register for assisting an Ayurvedic dispensary and another in Bombay was struck off for his connection with an Ayurvedic college. Indigenous practitioners wanted an equal share of state recognition and funding, not to be treated as 'medical untouchables with whom no registered allopathic practitioner can professionally come into touch without losing his caste'.[31]

The attitude of the state and its senior medical officers towards indigenous medicine remained largely one of hostility. As indicated in chapter 3, practitioners of Western medicine had long had a pragmatic interest in the drugs employed by *vaids* and *hakims*, and this was sustained into the late nineteenth century by such works as E. J. Waring's *Pharmacopoeia of India* (1868) and U. C. Dutt's *Materia Medica of the Hindus* (1877). In common with the revival of interest in Indian handicrafts, the continued investigation of medicinal drugs was stimulated by local and international exhibitions. George Watt and T. N. Mukherjee gathered a large number of samples for the Calcutta International Exhibition of 1883–4; collecting nearly 1,000 specimens for the exhibition prompted Mohideen Sheriff in Madras to publish his own *Materia Medica* in 1891. The single most influential figure was Kani Lal Dey, who had risen from the ranks of sub-assistant surgeons to become Professor of Chemistry at Calcutta Medical College. When the Indian Medical Congress met for the first time in Calcutta in 1894 several papers in the pharmacology section touched on the importance of 'indigenous drugs'. Watt listed the names and properties of a number of drugs and called for their further investigation, while Dey, as section president, urged the adoption of practical measures, including trials with selected drugs at hospitals and dispensaries, the cultivation of medicinal plants at government farms, and the setting up of a 'drug emporium' to provide a regular supply of indigenous drugs. Dey believed that the advancement of 'Indian pharmacology' by such means was 'in complete sympathy' with the 'development of commerce, medicine, and science in this country'

---

[30] Abdul Halim Sharar, *Lucknow: The Last Phase of an Oriental Culture* (London, 1975), p. 98.
[31] *The Report of the Committee on the Indigenous Systems of Medicine, Madras* (Madras, 1923), p. 25.

and would bring India's 'marvellously liberal provision of curative and remedial agents within the reach of the teeming millions of this Empire'.[32]

The Congress passed resolutions supporting Watt and Dey's call for the 'extended use of indigenous drugs' and, stung by criticisms of its neglect of medical research, the government took the resolutions seriously. Although the Surgeon-General was sceptical, arguing that the people of India used indigenous drugs 'simply because they could not get anything better', the Home Department set up an Indigenous Drugs Committee to investigate their use.[33] The committee, which initially included Watt and Dey, lasted until 1920, but had lost momentum well before the publication of its second report in 1909 and yielded few practical results. The investigation proved technically far more complicated than originally anticipated; it also encountered antagonism from European physicians, many of whom did not 'feel justified in experimenting with unknown and doubtful drugs' when they already had 'others of known and well-established efficacy available'.[34]

Investigations into the properties of indigenous drugs continued intermittently, mainly as a sop to the nationalist press and Indian members of the legislative councils. They were usually delegated (as an unrewarding task) to Indian medical officers and, by focusing on specific drugs rather than on the indigenous systems as a whole, they provided a 'scientific' basis for marginalising or refuting claims made for the superior efficacy of indigenous medicine. An investigation begun in Madras in 1918 involved clinical trials, but predictably upheld the view that virtually all of India's potentially useful drugs had already been identified.[35] In 1930 **Ram Nath Chopra**, IMS, of the Calcutta School of Tropical Medicine, was appointed to head another inquiry. He produced a substantial report, which supported the view that several Indian drugs might usefully be added to the Western pharmacopoeia but without reviving and officially recognising the indigenous systems.[36] In 1947, on the eve of Independence, Chopra was appointed to head a more wide-ranging committee to consider the state's relationship with the systems of indigenous medicine and to suggest ways of enhancing their usefulness to the public. The tone of the report was sympathetic, but the overall message was to confirm the superiority of allopathic medicine, especially with regard to public health.[37]

Few state medical officers showed any sympathetic interest in Ayurvedic and

---

[32] Girindranath Mukhopadhyaya, *History of Indian Medicine from the Earliest Ages to the Present Time*, 1 (Calcutta, 1923), p. 37.

[33] *Report of Proceedings of the Central Indigenous Drugs Committee of India*, 1 (Calcutta, 1899), appendix, pp. 1–2; Home (Medical), 15–18, December 1895, NAI.   [34] *Drugs Committee*, 1, p. 226.

[35] M. C. Koman, *Report on the Investigation of Indigenous Drugs* (Madras, 1921).

[36] R. N. Chopra, *Indigenous Drugs of India: Their Medical and Economic Aspects* (Calcutta, 1933), p. 17.

[37] *Report of the Committee on Indigenous Systems of Medicine* (New Delhi, 1948).

Unani medicine. One apparent exception was Sir Pardy Lukis, Director-General of the IMS. In a debate in the Imperial Legislature in 1916, responding to a resolution calling on the government to 'investigate the possibility of placing the ancient and indigenous systems of medicine on a scientific basis and increasing their usefulness', Lukis declared that in thirty-six years in India he had given much thought to this question, and the longer he remained in the country the more he was 'convinced ... that many of the empirical methods of treatment adopted by the *vaids* and *hakims*' were 'of the greatest value'. He rejected the view that all *vaids* and *hakims* were quacks and saw no great division between the medicine of West and East, except that the former had advanced whereas the latter had stood still. He recognised that, since 90 per cent of the population lived in the countryside and had little access to Western medicine, it made sense to support indigenous medicine as far as possible so that the basic health needs of the people could be met.[38]

Lukis's views were widely welcomed as an endorsement of the revitalisation movement and were often cited in subsequent debates. But, cautious though his words actually were, his opinions were not widely shared in the IMS. However, the central and provincial governments were anxious not to alienate moderate Hindu and Muslim opinion at a time of growing political unrest, and the contempt with which many European physicians and ICS administrators personally viewed indigenous medicine was often muted in their public statements. Such evasive tactics were, however, increasingly difficult to maintain. In order to retain the support of their moderate allies during the First World War, the British responded to nationalist demands for Home Rule in 1917 by promising Indians a degree of 'responsible government'. This was watered down in the Government of India Act of 1919 to a system of 'dyarchy' by which power in the provinces was divided between 'reserved' departments under executive councillors and 'transferred' departments in the hands of elected ministers: the former included the most vital portfolios, such as revenue and police, whereas ministers were handed under-funded, second-rank departments such as education, health and agriculture. This had the effect, however, of making medical expenditure and health policy subject to a greater degree of political scrutiny. In 1918 and 1920 the Indian National Congress passed resolutions stating the 'undeniable claims to usefulness' of the Ayurvedic and Unani systems and calling for the establishment of 'schools, colleges and hospitals for instruction and treatment in accordance with the indigenous systems'.[39] Although the

---

[38] *Proceedings of the Indian Legislative Council, 1916–17*, 65, pp. 337–49.

[39] Roger Jeffrey, 'Doctors and Congress: The Role of Medical Men and Medical Politics in Indian Nationalism', in Mike Shepperdson and Colin Simmons (eds.), *The Indian National Congress and the Political Economy of India, 1885–1985* (Aldershot, 1988), p. 161.

Congress boycotted the 1920 elections, members of the legislatures took up the cause of indigenous medicine and, in a further conciliatory move, the governments of Bengal and Madras agreed to set up committees of inquiry into the indigenous medical systems.

The most important of these was held in Madras, where a Committee on the Indigenous Systems of Medicine was appointed in October 1921. The committee was chaired by Mohammad Usman, a part-time *hakim*, but the key figure was the secretary, G. Srinivasa Murti, who concentrated the committee's attention on Ayurveda. A former allopathic medical officer and a Theosophist, Srinivasa Murti made a detailed and closely reasoned case for Ayurveda as a fully 'rational and scientific' system, self-sufficient in all respects apart from surgery. Claims for the scientific integrity of Ayurveda were supported by references to Brajendranath Seal, Gananath Sen, and Jagadis Chandra Bose, who, through the methods and instruments of modern science, had 'made the ancient teaching live once again in our minds'. Srinivasa Murti appeared to reject any compromise with allopathy. The theory of *tridosha* (three elements) was, he argued, the core of Ayurvedic thought. It could not be equated with the humoral system of the Greeks and thereby dismissed as anachronistic; nor could Ayurveda simply be treated as a source from which useful drugs could be transferred to the allopathic pharmacopoeia. The philosophical and physiological system of Ayurveda had to be accepted in its entirety. However, when it came to practical proposals, Srinivasa Murti appeared more accommodating. Since allopathy reached barely 10 per cent of the population, the indigenous systems had to meet the health needs of the remaining 90 per cent, and since, Srinivasa Murti argued, India could not afford to have two systems, it was necessary for students of Western medicine to become acquainted with the indigenous systems, and vice versa, so that in the end, through a process of mutual assimilation and synthesis, there would emerge a 'unified and integrated whole'.[40]

The Usman Report was greeted with anger and dismay by many members of the Western medical establishment, who saw it as a nationalist diatribe that failed to offer any convincing evidence for the claims made for Ayurveda. Criticism of the 'unscientific' nature of the indigenous systems of medicine was, if anything, more intense in the 1920s than it had ever previously been.[41]

[40] Srinivasa Murti, 'Memorandum'; Shiv Visvanathan, *A Carnival for Science: Essays on Science, Technology and Development* (Delhi, 1997), pp. 136–43.
[41] 'Indigenous Systems of Medicine in India', *British Medical Journal*, 15 September 1923, p. 479; W. Burridge, *Indigenous Systems and Medical Science* (Allahabad, 1926); J. W. D. Megaw, 'The Ancient Systems of Medicine', *IMG* (December 1925), pp. 585–8.

Many members of the IMS believed that any accommodation with indigenous medicine would not only threaten their state funding and professional status, but also make them objects of ridicule, even ostracism, among their colleagues in Britain. For them syncretism was not a professionally or politically acceptable option. None the less, for political reasons of its own the Madras government overrode its medical advisers and accepted the Usman Report. With the influential backing of the Chief Minister, the Raja of Panagal, it sanctioned the setting up of a School of Indian Medicine, which was opened in November 1924 by the Governor of Madras with Srinivasa Murti as its Director.

The School aimed to give students 'such training . . . as will enable them to become competent practitioners of [the] Indian system of Medicine' as well as having 'a good working knowledge of the Western system'. In what was almost a return to the Calcutta Native Medical Institution of the 1820s, Ayurveda, Unani and Siddha were taught alongside modern anatomy, physiology and surgery. A postgraduate programme was begun and staff compiled textbooks and papers identifying 'correspondences' between the indigenous and allopathic systems. In 1932 the powers of the School, and its long-serving director, were further augmented by entrusting to them responsibility for the registration of all indigenous practitioners in the province.[42] But, like the Unani purists of Lucknow, the institution failed to satisfy the advocates of 'pure Ayurveda', who were incensed that the School should teach anything other than Sanskrit texts and deplored the creation of a 'hybrid system' of medicine that ignored what Srinivasa Murti himself had once proclaimed – the uniqueness and integrity of the *tridosha* system. On the other side, there was mounting criticism from the practitioners of allopathic medicine, who saw the School as producing poorly qualified doctors who had been fed a bewildering mixture of ideas from two irreconcilable medical systems and who graduated in less time than it took students to meet the requirements of the Madras Medical College. Usman was again asked to investigate, and his report in 1941 revealed many shortcomings and inconsistencies in the School of Indian Medicine. But the wartime government would not risk taking measures that might be seen to discriminate against indigenous medicine, and in 1947 the School was formally recognised as a College of Indian Medicine.[43] More through pragmatism than conviction, 'integrated' medicine had survived but without gaining the equality with Western medicine it had long sought.

[42] G. Srinivasa Murti, 'The Government Indian Medical School, Madras', *Handbook of the Indian Science Congress: Twenty-Seventh Session, 1940, Madras* (Madras, 1940), pp. 86–92.
[43] G. O. 5177, Education and Public Health, 10 December 1941, TNA.

SCIENCE UNDER SIEGE

Constitutional changes under the Government of India Act of 1919 had far-reaching effects on the organisation, funding and political complexion of late colonial science. The authors of the reforms, the Secretary of State for India, Edwin Montagu, and the Viceroy, Lord Chelmsford, supported the view of the Holland Commission that state support was essential if science were to play a more active part in India's economic and industrial development. In practice, however, again for reasons of political expediency, they sacrificed science as a centralised state enterprise in order to augment the powers entrusted to the provinces under dyarchy and so increase the attractiveness of the reform package.

Provincialisation weakened, in some instances destroyed, the all-India scientific services. The Indian Agricultural Service, which had heralded a new start to state engagement with agriculture when it was set up in 1906, was badly hit by wartime casualties among its European staff and by the requisition of many of its buildings. The implementation of the 1919 Government of India Act resulted in the transfer of agricultural research, education and development to the provinces, and the central government lost almost all responsibility for its direction and control. Instead of being the apex of agricultural and veterinary research, Pusa and Muktesar found themselves out of touch with the provinces that were originally intended to be their scientific fiefdoms. By 1924 the very future of the Agricultural Service was in doubt. The Royal Commission on Agriculture, appointed in 1926, emphasised in its report the vital importance of agricultural and veterinary research for the future of India's agrarian economy, but could see no place for the Agricultural Service, recommending that there should be no further recruitment and that the service be wound up. It proposed, instead, an Imperial Council of Agricultural Research to combine technical and administrative expertise and ensure the 'ordered advance' of agricultural research.[44] Agriculture was not the only victim. The Calcutta School of Tropical Medicine, originally intended as an all-India research centre, also found itself reduced to provincial status by the time it opened in 1921 and so subject to a host of unwelcomed provincial financial and political pressures.

Provincialisation aroused mixed responses. Many Indian, and some European, scientists held that Curzon's centralising policy gave too much control to the state bureaucracy and stifled individual initiative. An article in

---

[44] *Royal Commission on Agriculture in India: Report* (London, 1930), pp. 35–6, 626.

Nature in 1920 claimed that many of India's leading scientists opposed centralisation and preferred regional, non-governmental institutes. For original scientific investigation, 'little or no official control' was needed, and scientists should not be 'constantly called upon to furnish interim reports and programmes of work to an official chief, or to obtain his formal sanction before undertaking an investigation or publishing the results of their work'. Such formalities 'waste valuable time, lead to constant friction, and are altogether foreign to the spirit which should reign in all centres of creative scientific research'.[45] By contrast, in 1928 A. G. Clow (as an ICS officer arguably biased in favour of all-India services) regretted that Montagu and Chelmsford had favoured the provinces over the centre and in so doing stymied industrial and scientific progress without any compensating expansion of scientific and technical institutions in the provinces.[46]

It was this latter view that generally came to prevail in India by the 1930s and 1940s. Arguably, India needed both central institutes and scientific services as well as provincial institutes and university departments geared to research. In fact, it got an unsatisfactory mixture of the two that left many significant gaps in Indian research and technical training. In the case of agriculture, veterinary science and forestry, for example, universities failed, or were unable, to provide alternative research centres: Benares Hindu University was almost alone among Indian universities in having a department of agriculture before 1945. It became increasingly necessary, if vital research was to survive and develop, for the effects of the Montagu–Chelmsford reforms to be at least in part reversed and for new bodies like the Imperial Council of Agricultural Research to emerge to take responsibility for funding and coordinating research. It also became necessary for India's scientists to devise their own coordinating bodies and during the Second World War to build centralised institutions anew and thereby restore a more directional role to the centre. A product of political expediency, backed by understandable antipathy to state-run science, provincialisation held back the evolution of an effective all-India science policy for more than twenty years. No less critically, provincialisation was accompanied by severe retrenchment. In 1925, just as scientific institutions were recovering from wartime losses, the Indian Retrenchment Committee under Lord Inchcape directed extensive cuts in government expenditure that badly hit scientific and medical funding. Then, as the Depression struck India in 1930–1, a further round of cuts was ordered: science and medicine were subject to

---

[45] 'The Organisation of Scientific Work in India', *Nature*, 19 February 1920, p. 654.
[46] A. G. Clow, *The State and Industry* (Calcutta, 1928), pp. 17–25, 66–75.

further cuts and job losses. Retrenchment hit many areas of state activity, but the 'special selection of science for such slaughter showed how little the Government of India understood or valued such work'. In consequence, India faced the Second World War 'much worse equipped scientifically than would otherwise have been the case'.[47]

Under the 1919 Act the central Legislative Assembly gained the right to debate and vote on health budgets and funds for medical research though it was denied full financial control. The mounting political crisis of the late 1920s and early 1930s (in which the boycott of the Simon Commission in 1928 was followed by the Civil Disobedience Movement of 1930–3), combined with the Depression, created new opportunities for Indians in the legislatures to try to wrest further concessions from the government over health funding and medical research while espousing the cause of medical science in the universities. In a debate in the Legislative Assembly in February 1930, M. R. Jayakar of Bombay moved a resolution calling for the long-mooted Central Research Institute for medicine to be located in a major city (Bombay or Calcutta) rather than at Kasauli or Dehra Dun. 'I hope', he said, 'the Government will realise that the days of isolated hilltops are gone for ever.' He also called for changes in the composition of the Governing Body of the Indian Research Fund Association (IRFA) and in the appointments board for the Medical Research Department, moves designed to strengthen Indian representation and control. These demands reflected deep dissatisfaction in the Indian medical profession and university medical faculties in Bombay and elsewhere at their exclusion from state-funded research and institutional posts.[48]

Senior IMS officers privately protested that such moves would be disastrous for the future of medical research in India: 'scientific' direction would be replaced by 'political' control and European researchers would be pushed out in favour of politicians' protégés. None the less, mounting political pressure in India forced the government to adopt a conciliatory response. In February 1930 plans for a central medical institute were abandoned and the composition of IRFA's Governing Body (which had previously been drawn almost exclusively from the IMS and ICS) was revised to give Indian non-officials an equal share of membership. The Depression brought further pressure to bear. The Retrenchment Committee, set up by the government but with strong Indian membership, proposed swingeing cuts in the medical establishment, including the abolition of the Director-General of the IMS and the Government of

---

[47] 'Science and Progress in India', *Nature*, 5 May 1945, pp. 526–7.
[48] *Legislative Assembly Debates, 1930*, 1, 8 February 1930, pp. 543–8; ibid., 11 February 1930, pp. 576–89.

India's Health Commissioner. These proposals were successfully resisted by the government but only by conceding a massive reduction in IRFA's annual grant from Rs 7 lakhs to 1.5 lakhs.[49] The government was forced to accept that in future medical research committees could no longer be monopolised by officials but would have to include a substantial number of non-official, Indian members.[50] In March 1938 IRFA became formally independent and, despite a one-off grant for malaria work in 1936, government grants to IRFA were not fully restored until 1944.[51]

Despite the deep animosity aroused by these manoeuvres, there was more to this than a simple racial or colonial/national divide. In 1931, when the zoological and botanical surveys were threatened with virtual closure, Indian as well as European voices were raised in protest. Raman observed how in recent decades growing numbers of young Indians had become interested in scientific research, often in subjects that had a direct bearing on India's future welfare. Some had found an opportunity both to follow a scientific avocation and to serve their country by joining the state services and institutes. This, Raman declared, 'not only brought credit to India' but promised 'to be of real national importance'. He bitterly regretted that in India's higher administrative circles there seemed to exist 'an insufficient appreciation of the close relationship between scientific research and national welfare'.[52]

But the threat to Indian science did not come from the philistinism of India's administrators alone; it came, too, from within the nationalist camp. For as long as liberal moderates and modernisers had command of the nationalist movement, science seemed secure. Ray's patriotic address to the Indian Science Congress in 1920 exemplified confidence in the complementarity of science and nationalism as first projected by Mahendralal Sircar fifty years earlier. But 1920 was also the year in which Gandhi seized control of that other Congress, the Indian National Congress, and converted it to a programme of non-cooperation. Gandhi was not entirely opposed to modern science: he occasionally professed his 'admiration' for it and his ideas of sanitation and diet were substantially informed by it.[53] But his damning critique of modern civilisation, of factories, railways and machinery, left scant room for Indian, as well as imperial, science and technology. Gandhi's call for Indians to renounce

[49] Education, Health and Lands (Health), 14/32, 1932, NAI.
[50] J. E. Turner to Director, Rockefeller Foundation, 14 March 1932, RAC.
[51] C. G. Pandit and K. Someswara Rao, *Indian Research Fund Association and Indian Council of Medical Research, 1911–61: Fifty Years of Progress* (Delhi, 1961), pp. 6, 10, 13.
[52] *MR* (September 1931), pp. 350–1.
[53] Gandhi to Nehru, 5 October 1945, in Jawaharlal Nehru (ed.), *A Bunch of Old Letters* (2nd edition, Bombay 1960), p. 506.

government titles, boycott state schools and colleges and take up the *charka* made India's leading scientists with their knighthoods and formal dress look self-centred, even anachronistic. Their commitment to cosmopolitan science seemed, in the face of Gandhi's onslaught in *Hind Swaraj*, inappropriate, even unpatriotic. Certainly, many Indian scientists continued to see themselves as sincere nationalists, working (as Raman's remarks indicate), even in the state services, for the nation's future welfare. Some, like Ray and the physicist **Meghnad Saha**, persuaded their students not to join the ICS, but to serve India through science. But, once a whiff of treachery hung over science and academic life in general, it was not easily dispelled.[54] The 1920s and early 1930s witnessed a 'certain amount of anti-intellectualism and anti-scientific attitudes',[55] and science at first found few defenders.

This reaction against modern science was not due to Gandhi alone. In the wake of the First World War, some Indian scientists had themselves begun to shrink from modernity. Pramatha Nath Bose, once a leading advocate of modern science, technical education and Swadeshi industry, by 1921 had come to take a much gloomier view. This might have been a consequence of advancing age, but repulsion at the mechanised barbarity of the First World War and the social changes he saw occurring in Bengal also influenced his antipathy. In seeking to explain the recent decline of the Bengali race, Bose blamed, among other things, the abandonment of traditional diets, hygiene practices and lifestyles, the adoption of allopathy, and the younger generation's fascination with speed, machines and the allure of modernity. He warned his countrymen that 'world-wide degeneration, physical and moral,' would be the 'bitter fruit of modern progress'. India should retire from the 'immature' civilisation of the West and return to a 'simple life'.[56] Ray, ever the ascetic, also drifted away from experimental science as he grew more conscious of rural poverty and turned to the *charka*, flood relief, and measures to revive the dying villages of Bengal. He, too, lamented the physical and moral decline of the Bengalis and in his old age deplored such evils of modern life as tea and motor cars.[57]

INSTITUTIONAL SCIENCE

Despite the political turmoil of the inter-war period, science in India underwent profound institutional changes. These were most evident in the

---

[54] Edward Shils, 'The Academic Profession in India', in Edmund Leach and S. N. Mukherjee (eds.), *Elites in South Asia* (Cambridge, 1970), p. 196.
[55] U. Sen et al., *Scientific Research in India Universities* (New Delhi, 1965), p. 2.
[56] Pramatha Nath Bose, *Survival of Hindu Civilisation, Part II* (Calcutta, 1921); *National Education and Modern Progress* (Calcutta, 1921), pp. 23–4.
[57] Prafulla Chandra Ray, *Life and Experiences of a Bengali Chemist*, II (Calcutta, 1935).

Table 6.1 *The founding of science departments at selected Indian universities*[58]

University	Botany	Chemistry	Geology	Mathematics	Physics	Zoology
Aligarh (1920)	1920	1920	1950	1920	1920	1932
Allahabad (1887)	1922	1922	–	1922	1922	1922
Andhra (1926)	1945	1932	1941	1932	1932	1947
Annamalai (1929)	1931	1929	1953	1929	1929	1931
Benares (1916)	1919	917	1921	1917	1917	1919
Calcutta (1857)	1918	1915	1928	1916	1916	1919
Dacca (1921)	n. a.	1921	n. a.	n. a.	1921	n. a.
Lucknow (1921)	1921	1921	1951	1921	1921	1921
Madras (1857)	1933	1933	1952	1927	1952	1933
Osmania (1918)	1924	1918	1945	1918	1928	1924

universities. At the outbreak of war in 1914 India had only five universities – Calcutta, Bombay and Madras, dating from 1857, with the University of Punjab at Lahore (1882) and the University of Allahabad (1887). Between 1916 and 1930 not only did the established universities, led by Calcutta, create new departments for such subjects as chemistry, mathematics and physics, but a substantial group of new universities came into being with a strong science orientation (table 6.1).

As the table indicates, the expansion in science departments was most marked in chemistry, physics and mathematics, the subjects that had most dramatically captured India's scientific imagination in the 1890s. It is indicative of this trend that, at Calcutta's Science College in 1921, 48 of the 174 science postgraduate students were studying physics, applied mathematics had 40 students, and chemistry and applied chemistry together 62, but in the department of physiology there were only 12 students, in geology 9, in zoology 7 and in

[58] Based on Sen et al., *Scientific Research*, pp. 75–85.

botany only 4.⁵⁹ In a significant inversion of the priorities of nineteenth-century colonial science, in the national science of the twentieth century botany, geology and zoology were relatively slow to attract institutional funding and student numbers. In several universities, geology departments were not founded until after Independence. Calcutta, an innovator as well as a pioneer in university expansion in the sciences, opened departments of psychology (1916), anthropology (1920), physiology (1938) and statistics (1941). But several of the new universities taught, and offered research facilities in, a wide range of medical and technical subjects. Within a decade of opening, Benares Hindu University had departments of Ayurvedic medicine, metallurgy, mining, and electrical and mechanical engineering. In the south, whereas Madras University was slow to shake off its arts and languages orientation, Andhra University quickly set up departments of chemistry, mathematics and physics, followed by chemical technology and applied physics. On the eve of Partition in 1947, there were a dozen or more departments in India of botany, chemistry, engineering, mathematics, physics and zoology, and more than two dozen university departments for medicine and allied subjects.

The opening of new departments of chemistry and physics created unprecedented opportunities for a new generation of academic scientists – in marked contrast to the 1890s. Bose, Ray and Raman were able to exact a kind of revenge for the lack of opportunities early in their own careers by nominating their protégés to key appointments, not just in Calcutta, but in Dacca, Benares, Allahabad and Lucknow as well. It was on their advice that the 27-year-old **Shanti Swarup Bhatnagar**, a Punjabi who had studied in Lahore and London, was appointed to the Benares chair of physical chemistry in 1921. In the same year, Satyendra Nath Bose, who later collaborated with Einstein on quantum statistics, left Calcutta to become a Reader in physics at the new University of Dacca. Like S. N. Bose, J. C. Ghosh, later head of Dacca's chemistry department, had been a student at Presidency College where he was taught, and inspired, by Ray and J. C. Bose. In 1922 Nilratan Dhar, another of Ray's Presidency College students, became Professor of Chemistry at Allahabad, and subsequently induced the young physicist Meghnad Saha, another former Calcutta student, to join him by taking the physics chair. Allahabad, like Dacca, became more than just a satellite of Calcutta's scientific establishment; it was also a centre for scientific research in its own right.⁶⁰ It is as well to note, though, that Indians were not the sole beneficiaries of the creation and expansion of science departments. At Dacca, for example, Professor

---

⁵⁹ Robert S. Anderson, *Building Scientific Institutions in India: Saha and Bhabha* (Montreal, 1975), p. 17.
⁶⁰ Ibid., pp. 7, 17.

E. R. Watson played a leading role in the development of organic chemistry through his work on synthetic dyestuffs.

The changing nature and growing stature of Indian science were reflected in other ways, too. Leaving aside A. C. Wadia, the Parsi engineer who was elected a Fellow of the Royal Society (FRS) in London in 1841 (and the dozens of Britons with Indian connections so honoured during the colonial period), ten Indians became FRSs between 1918 and 1947 (table 6.2) and in increasing numbers as Independence neared. Although the list is in some respects unrepresentative of the widening range of science in late-colonial India, it reflects a growing social and regional diversity. Between 1924 and 1944 the Royal Society elected as Fellows Meghnad Saha, from a low-caste family in east Bengal, S. S. Bhatnagar (a Kayastha) and Birbal Sahni from Punjab, Homi Bhabha, a Parsi from Bombay, C. V. Raman, a Tamil Brahmin, his nephew S. Chandrasekhar, and K. S. Krishnan, Raman's associate and successor as Palit Professor of Physics, also originally from south India. Of the ten elected between 1918 and 1945, only Jagadis Bose had ever belonged to the state scientific services. With the partial exceptions of Bose's plant physiology and Sahni's palaeobotany, the list reflects the intellectual pre-eminence of physics. Despite Ray, and later Bhatnagar, India's chemists made less of an international impression, and the biological sciences (still largely dominated by the state services) lagged far behind. Geology had to wait until 1957 when D. N. Wadia (formerly of the Geological Survey) became an FRS.

Recognition by the Royal Society was more than a matter of individual reputation. As indicated in previous chapters, the Royal Society had had a special role in overseeing science in India since the late eighteenth century. In the early twentieth century, India's FRSs symbolised the continuing importance of institutional and intellectual linkages between Indian scientists and the scientific establishment in Britain. This connection might be used to bring pressure to bear in the interest of science on the governments in Britain and India, but it also suggested the extent to which India's leading scientists continued to look to Britain as a model for their own scientific agendas and institutions. In practical terms the Royal Society (like some other British bodies) provided modest funds for scientific research in India in the inter-war period, as in the case of Saha, who received a grant of £1,500 from the Society in 1931. Many of India's leading scientists continued to send their most promising students to Britain, and they, in turn, established lasting ties to leading figures (and FRSs) such as F. G. Donnan at University College, whose Indian disciples included the 'Three Musketeers', S. S. Bhatnagar, J. C. Ghosh and J. N. Mukherjee. Friendships made between Indian students in Britain could in turn provide useful contacts

Table 6.2 *Indian Fellows of the Royal Society, 1918–47*[61]

Date elected		Principal field of research
1918	Srinivasa Ramanujan	Mathematics
1920	Jagadis Chandra Bose	Physics, plant physiology
1924	C. V. Raman	Physics
1927	Meghnad Saha	Astrophysics
1936	Birbal Sahni	Palaeobotany
1940	K. S. Krishnan	Physics
1941	Homi Bhabha	Nuclear physics
1943	S. S. Bhatnagar	Chemistry
1944	S. Chandrasekhar	Astrophysics
1945	P. C. Mahalanobis	Statistics

back in India. It might, however, be noted that Indian science was becoming, in some respects, more international in the inter-war years and less tied to London and Edinburgh. Germany became a desirable place to study physics and chemistry, and from the early 1920s the Rockefeller Foundation funded small numbers of Indian students to study public health in the United States. The Rockefeller Foundation also had a key role in funding the All-India Institute of Hygiene and Public Health, which opened in Calcutta in 1932, and in financing or advising on anti-malaria and anti-hookworm programmes. India's participation in the League of Nations in the 1920s and 1930s brought a further international dimension to late-colonial health policy, but, like the Rockefeller Foundation, these organisations tended to work through existing administrative channels and so to reinforce the authority of the colonial medical establishment rather than subvert it.

Another indicator of the growing complexity and status of Indian science was the development of scientific associations. In 1910 India had only ten (including three for medical sciences and six for engineering and technology). By 1920 the number had grown to sixteen, by 1930 to twenty-seven, and by 1940 to thirty-eight.[62] Among the most significant of these bodies were the Mathematical Society (1911), the Indian Botanical Society (1920) and the Indian Chemical Society (1924), all three representing the emergence of university-based scientific disciplines. The creation of the Botanical Society, largely through the efforts of Shiv Ram Kashyap at Lahore, was a significant

[61] A. V. Hill, *The Ethical Dilemma of Science and Other Writings* (New York, 1960), p. 378.
[62] A. Rahman, R. N. Bhargava, M. A. Qureshi and Sudarshan Pruthi, *Science and Technology in India* (New Delhi, 1973), p. 16.

indication of how a field once dominated by Europeans and the services was now being appropriated by Indian academics. The establishment of the Indian Chemical Society, first proposed by Watson and Ray, was a similar landmark. Ray believed that India now needed its own society and a journal in which to publish the increasing volume of research papers by Indian chemists. He saw the formation of the Society, along with the several university posts recently occupied by his pupils, as fulfilling his dream forty years earlier in Edinburgh that India would contribute its share to 'the world's stock of scientific knowledge'. At the start of 1930, the Society had 460 members, with branches in Calcutta, Lahore, Bombay and Madras; but membership and journal subscriptions were hit by the onset of the Depression: by the end of the year there were only 377 members and 102 subscribers.[63] At the all-India level, the Indian Science Congress steadily established itself as the principal forum for Indian science. When the Congress first met in Calcutta in 1914 it had only six sections. By 1938 this had risen to thirteen, with agriculture, entomology, physiology and psychology among the newcomers. In that year, over 800 papers were presented compared with 35 in 1914, and, to celebrate the Congress's silver jubilee, it met in joint session with the British Association for the Advancement of Science, an occasion that further symbolised metropolitan recognition of Indian science.[64] Indians were also by the mid-1930s beginning to take charge of many of India's leading scientific institutions and research institutes, including the Haffkine Institute in Bombay in 1932 and Calcutta's Botanic Gardens in 1937. Against this trend, the All-India Institute of Hygiene and Public Health in Calcutta had an American, John B. Grant, loaned by the Rockefeller Foundation as its Director from 1939 until the end of the war. But in general, in science, as in other fields, it seemed that advancing Indianisation was beginning to spell the end of colonial rule.

And yet the picture was far from rosy. In the inter-war period, Indian scientists struggled hard both to gain recognition from their own government and to create organisations that would provide them with a common platform. In this they were often plagued by their own internal divisions. When Saha, as President of the Indian Science Congress in 1934, proposed the creation of a national scientific academy, Raman, now Director of the Indian Institute of Science, responded by setting up his own Indian Academy of Sciences based at Bangalore. As a compromise, a National Institute of Sciences of India was formed in 1935, with its headquarters in Calcutta, aiming to take on the role of the Royal Society in London or the National Academy of Sciences in

[63] *Report of the Indian Chemical Society for 1930* (Calcutta, 1930), pp. 1–3, 12, 15–16.
[64] Prashad (ed.), *Science*, p. xxvi.

Washington. Its objectives were to promote scientific knowledge in India, including its practical application to problems of national welfare; to 'effect coordination' between scientific institutes, societies, and government scientific departments and services; and to represent the interests of Indian scientists nationally and internationally.[65] Although this was an important step, the National Institute of Sciences proved less effective than had been hoped by its founders either at coordinating national science or in representing its interests to the Government of India and the world at large.

The government was not ready to listen to the voice of 'non-official' science, even when the crisis of the Second World War struck India. Representations were made to Delhi on behalf of scientists in Britain by the Secretary of State for India in 1941, but the Government of India remained unresponsive and India was denied participation in meetings that enabled Britain, the Dominions and the United States to pool their scientific and technological expertise. A Board of Scientific and Industrial Research was set up, rather half-heartedly, in 1940 with Bhatnagar as its Director; this was transformed into a Council of Scientific and Industrial Research in April 1942. An Industrial Research Fund was formed with an annual grant set at Rs 10 lakhs for five years. In 1943 the Council of Scientific and Industrial Research appointed a Research Planning Committee to survey the existing facilities for scientific and industrial research in India, which called for a major expansion in training scholarships and national research laboratories. But further substantial moves awaited the end of the war and then the British departure from India.

In 1944, following his visit to India on behalf of the Royal Society, A. V. Hill, the Nobel prize-winning physiologist, observed that many Indian scientists talked 'bitterly or sadly' about the poverty of their institutions, their low salaries, the under-staffing of laboratories, the shortage of equipment, and 'the lack of contact with the outside scientific world'. Hill stressed their positive achievements and the remarkable advances that had been made over the previous twenty-five years. However it was measured, he argued – by the number of scientific societies and journals, by the number of science graduates and FRSs, or 'by the contributions which India has begun to make to world science' – the progress of Indian science was 'a very real thing'.[66]

None the less, Hill's own report, published in 1944, showed that all was not well. The war had left India's scientists 'sorely cut off . . . from intellectual contacts with the rest of the world', and as a result the scientific and technical

[65] Ibid., pp. xxiv–xxv.   [66] Hill, *Dilemma*, pp. 373–4.

resources of India had 'not been utilised, or developed for war purposes to anything like the same degree as those of the other major countries'.[67] He found from his own tour that the standard in Indian medical colleges was 'on the whole unduly low' and there was 'little significant research'. India had a high reputation for tropical medicine, but the research had mainly been done in specialist institutes, leaving departments in universities and medical colleges bereft of funds and the stimulus of research.[68] In other scientific fields, too, Hill found a similar record of lost opportunities and deficient funding. Excellent work was being done in agricultural research, for instance, but only about Rs 70 lakhs a year was spent on it; India was 'not so poor that a considerably greater effort could not be devoted to research in its greatest industry and its most important interest'. Hill was struck by the way in which Indians saw national progress in terms mainly of industry and the ways in which physics, chemistry and engineering could serve its needs. Personally, he believed that the biological sciences had an equally important contribution to make, especially to agriculture, fishing, forestry and animal husbandry. But retrenchment had left the botanical and zoological surveys crippled and fishery research seriously under-funded.[69]

Finally, Hill turned to one of the main problems that had beset scientific research in India since 1920 – the lack of central direction and coordination. There had to be a common, all-India policy, he insisted. So long as the research work of the central government remained 'dispersed under a number of separate Departments or bodies, most of them having many other serious duties and preoccupations', there could be no realistic hope of arriving at a 'common plan' for science, and, if anything deserved to be common to India as a whole, 'surely scientific research can be among the first to claim that common interest'. He proposed a Central Organisation for Scientific Research, with six constituent boards (for medicine, agriculture, industry, surveys and resources, engineering and war), each with its own director, directly responsible to a central government minister for planning and development and charged with directing, initiating and coordinating research nationwide. This, he believed, would not only resolve many of the structural problems in the organisation of science in India, but also meet the complaints of India's non-service scientists, voiced through the National Institute of Sciences, that they were not being given the chance to contribute to the war effort and public welfare, had little influence with the government, and were being starved of research funds and opportunities.[70] Significantly, as a further indication of how metropolitan

[67] A. V. Hill, *Scientific Research in India* (London, 1944), pp. 2–6.   [68] Ibid., pp. 7–13.
[69] Hill, *Dilemma*, p. 376.   [70] Hill, *Research*, pp. 29–31.

initiatives still counted, Hill's proposals provided one of the bases for the post-war, post-Independence reorganisation of scientific research under the Government of India.

TROPICAL MEDICINE AND PUBLIC HEALTH

To a considerable degree, the medical research of the inter-war period continued to be situated within the paradigm of tropical medicine that had been so central to overcoming the epidemiological crisis of the 1890s and in reviving IMS prestige. The prospect of research into tropical diseases continued to be the bait offered to IMS recruits and it dominated the funding and research strategy of IRFA and, from 1921, the Calcutta School of Tropical Medicine. The dominance of the tropical paradigm was reflected in the large number of works on tropical veterinary science, tropical surgery, even tropical midwifery, published in India in the 1920s and 1930s and in IMS participation in wider medical and sanitary activities sponsored by the League of Nations and the Far Eastern Association for Tropical Medicine, which held its annual session in Calcutta in 1927. Indian medical researchers, too, worked extensively within the topical paradigm. Upendranath Brahmachari, who investigated the use of antimony in the chemotherapy of kala-azar in the 1920s, saw his research as a major contribution to combating one of India's principal tropical diseases and hence of parallel importance to Ross's malaria work twenty years earlier.[71] R. N. Chopra, Professor of Physiology at the Calcutta School, combined an interest in Indian pharmacology with 'studying problems of practical utility to professional medical practice in the tropics'. In 1936 he published a textbook on tropical medicine that differentiated between the therapeutic needs of people in tropical and in temperate countries.[72]

As mortality from smallpox, cholera and plague began to fall (in part reflecting the decline of famine as well as improvements in public health and medical treatment), so malaria assumed greater prominence as the leading cause of Indian mortality (estimated at 1 million deaths a year) and the tropical disease most urgently requiring investigation. Under the auspices of IRFA and the Malaria Institute (formed in 1927 at Kasauli), detailed malaria surveys were carried out, along with studies of India's various species of anopheles mosquitoes, their distribution, breeding habits and modes of control. Research

---

[71] Upendranath Brahmachari, *Gleanings from My Researches, I: Kala-Azar, Its Chemotherapy* (Calcutta, 1940).

[72] R. N. Chopra, *Handbook of Tropical Therapeutics* (Calcutta, 1936); B. Mukherji, 'Sir Ram Nath Chopra', *Biographical Memoirs of Fellows of the Indian National Science Academy*, IV (New Delhi, 1986), pp. 171–85.

on malaria carried tropical medicine to its highest levels of sophistication, combining the skills of epidemiology, entomology and protozoology. The Malaria Institute was praised nationally and internationally for work 'of the highest quality' and by 1946 was said to have established 'a reputation for malaria research . . . probably unequalled by any other single organisation in the world'.[73] But the Malaria Institute also demonstrated the limitations of tropical medicine as it had come to be institutionalised in India by the 1920s and 1930s. It focused on the research of a small number of mainly European IMS officers and was based in a remote hill-station. Not surprisingly, Kasauli was one of the main targets of attack by Indian politicians in 1930.[74] Whatever the ultimate scientific value of its work, it seemed remote from the practical, everyday difficulties of malaria control in Indian cities, towns and villages, where funding and basic skills and facilities were urgently needed but were in short supply. It was sometimes stated, 'with a considerable degree of truth', according to one IMS officer, that scientific research and the knowledge of disease in India had 'far outstripped its practical application'.[75] Whether from lack of resources or from provincialisation, there was a chronic failure to develop an all-India anti-malaria policy.[76] The capital, Delhi, received privileged treatment in the 1930s and 1940s, but even there effective malaria control remained stubbornly elusive. By 1947, experiments with DDT spraying seemed to offer the best hope of eradicating malaria-carrying mosquitoes and engendered a new confidence in the ability at last to defeat India's most deadly disease.[77]

Many of the most important advances in medicine and public health in the inter-war period lay outside the immediate field of tropical medicine and suggested that poverty and social deprivation were more important than a tropical location in determining Indian ill health. The continued growth of the women's medical movement directed new attention to reproductive health and the diseases of women. In the late 1920s Margaret Balfour of the Women's Medical Service showed, through an analysis of hospital returns, that maternal mortality in India was much higher than in most Western countries. Anaemia, eclampsia and osteomalacia, conditions that were relatively rare in

---

[73] *Report of the Health Survey and Development Committee*, 1 (Delhi, 1946), p. 18.

[74] *Proceedings of the Conference Held at Simla on the 21st and 22nd July 1930 to Discuss the Location of the Proposed Central Medical Research Institute and Other Matters* (Simla, 1931).

[75] *Annual Report of the All-India Institute of Hygiene and Public Health, Calcutta, 1934* (Calcutta, 1935), pp. 13–14.

[76] V. R. Muraleedharan and D. Veeraraghavan, 'Anti-Malarial Policy in the Madras Presidency: An Overview of the Early Decades of the Twentieth Century', *MH*, 34 (1992), pp. 290–305.

[77] Education, Health and Lands (Health), 23–1, 1939, NAI; Paul F. Russell, *Man's Mastery of Malaria* (London, 1955), pp. 168–70.

the West, were leading causes of maternal mortality in India partly because of the effects of poor nutrition, along, it was believed, with social and cultural factors like purdah.[78] In 1936–7, IRFA funded an investigation into maternal mortality in India and produced important findings, but when the first All-India Conference of Medical Women met at Delhi in 1938 it was pointed out how neglected conditions affecting maternal health were in terms of state-funded medical research as a whole, despite their obvious importance to the millions of India's women.[79] It was also striking that, when the All-India Institute of Hygiene and Public Health opened in Calcutta in 1932, the original plan to include a department of maternity and child welfare was shelved because of the financial crisis. For four years a smaller unit functioned there, but only through support from the Dufferin Fund. That there were any advances in the investigation of maternal and child health owed much to women doctors working outside the state section, for municipalities or voluntary organisations like the Indian Red Cross and the Ramakrishna Mission.

Industrial health, too, was coming under closer scrutiny. Before 1914 there had been a tendency to argue that factory conditions were healthier than those of the rural population from which workers were drawn. Factory workers lived, by some accounts, 'in a state of comfort ... almost unknown to ordinary labourers of the same class'.[80] But by the 1920s such claims were increasingly untenable. The growth of the industrial workforce (reaching 1 million by 1915) and the introduction of labour legislation forced belated recognition of the health hazards to which industrial workers were exposed as a result of factory accidents, poor nutrition, long working hours, insanitary conditions and inadequate medical attention. Diseases such as anaemia among plantation labourers and tuberculosis among factory hands were closely related to diet and working conditions. But medical research in these fields remained piece-meal and unsystematic.[81] It is again significant that the All-India Institute of Hygiene and Public Health, despite its location in Calcutta, one of the country's largest industrial centres, lacked a department of industrial or occupational health. It was not until the Second World War,

---

[78] Margaret Balfour, 'Maternal Mortality in Childbirth in India', *Journal of the Association of Medical Women in India*, 16 (1928), pp. 5–25.

[79] 'Extracts from the Report of the Public Health Commissioner for 1937', *Journal of the Association of Medical Women in India*, 27 (1939), pp. 277–9; M. I. Neal, 'Anaemia: A Challenge', *Journal of the Association of Medical Women in India*, 26 (1938), p. 16.

[80] E. A. Bradbury, Serampore, to Magistrate, Hooghly, 5 April 1879, Legislative, 165, April 1881, NAI.

[81] Margaret Balfour, 'Maternity Conditions and Anaemia in the Assam Tea Gardens', *Journal of the Association of Medical Women in India*, 21 (1933), pp. 28–38; A. Lankester, *A Report on Tuberculosis in India* (Simla, 1915), pp. 84–5.

when the demand for increased industrial output coincided with plans for post-war reconstruction, that arguments for improving workers' health were fully spelled out. 'It cannot be too strongly emphasised', declared the head of the Industrial Health Advisory Committee in 1945, in calling for a medical inspectorate for factories and mines, 'that the efficiency of industry and prosperity of a nation are intimately linked with the well-being and health of its industrial population.'[82]

Along with malaria, malnutrition was one of the principal sites of medical enquiry in late-colonial India. Before 1900, apart from a passing interest in beri-beri and 'famine foods', there had been no systematic, scientific investigation of Indian diets, though in both British and Indian discourse the apparent deficiency of rice diets was repeatedly alluded to. Work by D. McCay, Professor of Physiology at Calcutta Medical College, on jail diets in Bengal and the United Provinces in 1910–12 was the first to address scientifically the differences between rice and wheat diets and their physiological effects, thus paving the way for more sophisticated studies.[83] After the First World War, rapid advances in the knowledge of nutrition in the West helped **Robert McCarrison**, IMS, with support from IRFA, to establish a nutritional research programme at Coonoor. Following up earlier research on beri-beri in Java and Malaya, McCarrison demonstrated the differential effects of rice and wheat diets on laboratory animals, and showed the role of milled rice (itself an example of the widespread impact technological change could have) in causing poor physique and deficiency disease.[84] Although hampered by lack of funds (the beri-beri enquiry, axed in 1923 on financial grounds, was restored two years later), McCarrison was able to gain significant recognition for the importance of his work. He made a lasting impression on members of the Royal Commission on Agriculture when they visited Coonoor in 1926, convincing them that '[o]f all the disabilities from which the masses in India suffer, malnutrition is perhaps the chief', and stressing the link between nutritional science and improvements in Indian agriculture.[85] A decade later, when the Commission's chairman, Lord Linlithgow, became Viceroy of India he showed a personal interest in nutrition, pushing it to the top of the research agenda. In 1936 a Nutrition Advisory Committee was established and roughly a tenth of IRFA's annual grants went to fund nutrition research at Coonoor and Calcutta.

[82] *Indian Industrial Health Advisory Committee, 1945–6*, OIOC, V/26/840/16, p. 5.
[83] D. McCay, *Investigation into the Jail Dietaries of the United Provinces* (Calcutta, 1912).
[84] Robert McCarrison and Roland V. Norris, 'The Relationship of Rice to Beri-Beri in India', *Indian Medical Research Memoirs*, no. 2 (1924).
[85] *Royal Commission on Agriculture in India, I: Evidence* (London, 1927), pp. 95–8; N. Gangulee, *The Indian Peasant and His Environment* (London, 1935), p. 118.

Viceregal patronage still counted in science policy, though so, too, did the League of Nations, which showed a keen interest in nutrition from the early 1930s.[86]

For all its wider implications, McCarrison's work was confined to the laboratory. It was not until the arrival of his (non-IMS) successor at Coonoor, **W. R. Aykroyd**, that nutrition research came closer both to the lives of the Indian people and to state policy. As Director of the Nutrition Research Laboratories from 1935 to 1946, Aykroyd and his Indian co-workers conducted a series of careful surveys of the diets of school children, hostel students and villagers in south India. Over 50 rural and urban diet surveys had been carried out by 1939, and by 1947 there were 130. These covered a wide range of regions, social groups and institutions, and provide a significant example of how medicine had moved away from the remote essentialising of the nineteenth century to the more scientific study of human physiology, directed to meeting public needs and offering practical solutions to the widespread problem of malnutrition.[87] The findings of nutritional science also had implications for state policy. As a member of the Commission of Inquiry into the Bengal Famine in 1944, Aykroyd tried to push nutritional issues to the fore and to demonstrate their relevance to post-war reconstruction. Ten years of diet surveys had shown that malnutrition and hunger were by no means confined to famine episodes alone; even in 'normal times' a substantial proportion of the population did not get enough to eat. Indian critics sometimes suggested that nutritional deficiencies, like famines, were a result of misguided colonial policies and would disappear once freedom came. Aykroyd himself believed that freedom alone was not the solution. The state had a responsibility to cast aside outdated notions of laissez-faire and become actively involved in the production and distribution of food, and wartime experience had shown that it was both possible and necessary. Since defective nutrition had clearly been shown to be a major cause of Indian ill health, it was incumbent on any modern government to ensure that changes in diet were made in order to improve the people's health and physical well-being.[88]

For much of the nineteenth century, famine had been a major influence in shaping the character and priorities of colonial science, technology and medicine. Although famine returned, devastatingly, to Bengal in 1943–4, after 1908

---

[86] Education, Health and Lands (Health), 37–25, 1936; Education, Health and Lands, 52–27, 1938, NAI.

[87] W. R. Aykroyd, *Note on the Results of Diet Surveys in India* (New Delhi, 1939); S. L. Bhatia, 'Physiology', in Prashad (ed.), *Science*, pp. 575–630.

[88] W. R. Aykroyd, *Notes on Food and Nutrition Policy in India* (New Delhi, 1944), p. 1; *Famine Inquiry Commission: Final Report* (Madras, 1945).

the great destructive famines that had marked the second half of the previous century appeared to be over. But, as famine receded, population growth gave rise to unprecedented doubts, among Europeans especially, about the ability of modern science and technology to resolve India's economic and social problems. Perhaps reflecting an underlying loss of political self-confidence among the British, there was a new mood of pessimism that contrasted with the overweening confidence in the power of technology that had informed imperial attitudes in the late nineteenth century. Whereas writers such as MacGeorge forty years earlier had a profound belief in the ability of railways, irrigation canals and other major engineering works to transform India into a modern society and economy, now there was a growing awareness of the complex and vulnerable nature of the Indian environment and the cultural and political difficulties involved in trying to effect any change. Endemic malaria, chronic malnutrition, soil erosion and salinity were but some of the persistent difficulties that seemed to beset India and to which no easy scientific or technological solution could be found.[89]

In 1928 J. W. D. Megaw, IMS, one of a growing number of both Indian and European advocates of birth control, remarked: 'The people multiply like rabbits and die like flies: until they can be induced to restrict their rate of reproduction there is no hope of doing much good by medical relief and sanitation, as the population is very nearly up to the possible limit.'[90] By highlighting the scale and interrelatedness of India's agrarian difficulties, the report of the Royal Commission on Agriculture seemed to do as much as the census of 1931 (which showed an increase of 30 million in ten years) to nurture this dark mood. In 1935 Sir Stanley Reed, former editor of the *Times of India*, wondered how Indian agriculture could possibly keep pace with the 'colossal increase' in population. The question had become not how to improve, but how simply to maintain the existing standard of living.[91] Hill, whose visit to India coincided with the Bengal famine of 1943, took a more optimistic view of India's prospects and the part that science and technology could play; but he, too, believed that unless India curbed its birth rate it could make no lasting improvement in health, nutrition and social welfare. He reckoned that India would need a threefold increase in food production over the next thirty years; this could be done, but it was a 'tremendous' task, requiring a 'very great national effort' and the 'fullest use of modern scientific knowledge and methods'.[92]

In the meantime, the prospects, as far as health was concerned, looked

---

[89] *Report of the Punjab Erosion Committee, 1932* (Lahore, 1934).
[90] Megaw to W. S. Carter, 29 October 1928, RAC.   [91] 'Foreword' to Gangulee, *Peasant*, p. xxvi.
[92] Hill, *Dilemma*, pp. 337–46.

bleak. In October 1943 the Government of India set up a committee, chaired by Sir Joseph Bhore, to examine existing health provision in India and make recommendations for 'post-war development'. Two-thirds of the membership of the Health Survey and Development Committee was Indian (including the three women members), and among its numbers were many of the leading figures in medicine and public health in India at the time. In its report, published in 1946, the Bhore Committee criticised the poor record of public health in British India and pressed for a more effective and far-reaching system of public health to meet the needs of the Indian people. Health, it insisted, reflecting the mood of the Beveridge Report in Britain, should be available to all members of society, 'irrespective of their ability to pay for it'. At present, the death rate in British India was nearly twice that in England and Wales, infant mortality stood almost three times higher, and life expectancy at birth was less than half. The report was especially scathing about the poor state of maternal and infant health, estimating that maternal deaths alone numbered 200,000 a year. Any plan for improving public health must accordingly 'pay special attention to the development of measures for adequate health protection to mothers and children'. The report attributed India's massive health problems partly to the country's physical and sanitary environment, but also to severe deficiencies in the existing health services, which were 'altogether inadequate to meet the needs of the people'.[93] In the whole of British India, there were only 47,400 practitioners of modern medicine for a population of 300 million (one for every 6,300 people), and only 7,000 nurses and 5,000 trained midwives. The report concluded that the 'present low state' of public health was reflected in the 'wide prevalence of disease and the consequent high rates of mortality in the community as a whole and, in particular, among such vulnerable groups as children and women in the reproductive period'. It considered that at least half India's existing mortality was 'preventible and should therefore be prevented'.[94]

Not surprisingly, given its membership, the Bhore Committee did not reject Western medicine or press for radical alternatives — indeed, it was largely dismissive of Ayurvedic and Unani medicine, regarding public health as 'not within the purview of the indigenous systems of medicine as they obtain at present'.[95] It sought, instead, to expand Western health services and facilities and to develop them in the national interest. India needed more hospitals, dispensaries, doctors and nurses; it also needed more (and more relevant) medical research. Although the Bhore Committee praised institutions such as the

[93] *Report of the Health Survey and Development Committee*, I, pp. 8–11. [94] Ibid., I, pp. 11, 13.
[95] Ibid., II, p. 455.

Malaria Institute, it generally concluded that 'the achievements of the past have not been commensurate with the unequalled material available for research or with the pressing needs of the country'. The report supported the system of central funding under IRFA, but, reflecting years of Indian criticism, it believed the present organisation of medical research was far too 'narrow and restricted'; it regarded the 'almost complete absence' of research in Indian medical colleges and university medical faculties as an 'outstanding defect' that needed urgent rectification.[96]

SCIENCE FOR THE NATION

India by 1939 was among the eight most industrial countries in the world; it also had one of the largest scientific communities to be found anywhere outside Europe and North America. Although the political impact of these developments was in part muted by the rise of Gandhism and the relative indifference of the colonial regime, the period after 1900, and more especially after 1920, saw far-reaching developments in technology as well as science and medicine that had direct consequences for the life of the Indian people. Among the most momentous was the advent of electric power. In nineteenth-century India the main focus of hydraulic engineering was concentrated on the use of water resources for irrigation purposes; the potential for either navigation or power generation was largely ignored. Around 1900, although several large-scale irrigation schemes continued to be developed, hydro-electric projects were initiated that offered prospects for change in the countryside as well as in the cities. One of the earliest was the Sivasamudram scheme on the Cauvery, begun in Mysore in 1899 and in operation by 1902, which, in addition to supplying power to Bangalore, opened a new phase in the history of Indian mining by allowing deep workings in the Kolar goldfields. Other important schemes of the period included the Tata HEP project, completed in 1915, which used water stored above the Bhor Ghat to generate electricity for the city of Bombay, and the Pykara scheme in western Tamilnadu, begun in 1928, which opened up new opportunities for south Indian industry, especially the cotton-textile mills of Coimbatore.[97]

Transport and communications, too, underwent radical change and had far-reaching effects. Although the age of steam continued on India's railways, the internal combustion engine brought cars, buses and lorries to the streets of

---

[96] Ibid., pp. 409–10.
[97] W. C. Ash, 'Engineering', in Prashad (ed.), *Science*, pp. 484–8; Christopher John Baker, *An Indian Rural Economy, 1880–1955: The Tamilnad Countryside* (Oxford, 1984), p. 351.

Indian towns and cities and, aided by an expanding network of all-weather roads, brought new accessibility to the countryside. By the 1930s buses and lorries were plying many rural routes, where railways had never reached, ferrying cash crops to market, fostering the growth of a new generation of small-town entrepreneurs, and allowing policemen, politicians and mobile dispensaries to reach all but the remotest villages. On the eve of the Second World War, India had 175,000 motor vehicles.[98] Like hydro-electricity, the rapid rise of the internal combustion engine had a number of spin-off effects for Indian science as well as its industry. One was the fresh stimulus given to geology by the search for India's own oil resources; another was the growth of motor mechanics and truck drivers, many of whom (in Punjab at least) derived their skills from service with the armed forces during the Second World War.[99] By the 1930s, air travel, too, was beginning to facilitate international and domestic travel, with the first regular flights reaching India in the 1920s. At the start of the century, the lawyer and politician Motilal Nehru had been one of the first Indians in Allahabad to own a motor car. Thirty years later his son, Jawaharlal, flew to Europe to visit his dying wife and became one of the first Congress politicians to exploit air travel for political ends in the election campaign of 1936–7, with, it was said, 'considerable influence' on the outcome.[100] Communications, too, were advancing beyond the age of postal services and telegraphs inaugurated in the 1850s. Telephones now facilitated communication within and between towns. Radio broadcasting, begun on a small scale in 1926, was taken over by the state in 1930. The cinema ushered in a new era of mass entertainment and, like the railway and printing press before it, rapidly made an imported technology India's own.[101]

Technological change as much as scientific innovation bred a new confidence among India's scientific elite. It also fostered a more openly critical stance towards Gandhi. A defence of scientific modernity had been slow to surface. One of the first signs was a hostile reaction to Gandhi's remarks about the Bihar earthquake of 1934 as a divine response to India's failure to resolve the problem of untouchability. Several leading scientists expressed their incredulity and dismay at such a suggestion; as far as they were concerned the earthquake, soon followed by that at Quetta in 1935, demonstrated the urgent need for an Earthquake Commission and more research into Indian seismology.[102]

[98] L. S. S. O'Malley (ed.), *India and the West* (London, 1941), pp. 247–8.
[99] D. N. Wadia, 'Geology', in Prashad (ed.), *Science*, pp. 87, 103; Clive Dewey, 'Some Consequences of Military Expenditure in British India: The Case of the Upper Sind Sagar Doab, 1849–1947', in Clive Dewey (ed.), *Arrested Development in India: The Historical Dimension* (Riverdale, MD, 1988), pp. 141–2.
[100] O'Malley, *India*, p. 250. [101] Ibid., pp. 252–3.
[102] 'The Great Quetta Earthquake', *Science and Culture*, 1 (1935), pp. 63–7.

But only gradually, as the Gandhian ethos appeared to weaken within the Indian National Congress and as Subhas Chandra Bose and Jawaharlal Nehru seemed set on redirecting the Congress on a path more favourable to modern science and technology, did India's scientists begin to stir themselves and reassert their nationalist credentials.

In 1935, when Meghnad Saha and his associates launched the monthly journal *Science and Culture*, criticism was directed less at the colonial state than at the Gandhian vision of a low-technology society, based on hand-spinning and a traditional system of agriculture. Like many scientists, Saha did not believe that India could meet the challenge of the future by going back to the past. An editorial in the first issue of *Science and Culture* pointed to the 'considerable number' of India's leaders who appeared 'incapable of seeing the great and inevitable part which the new age of technic will play in India's destiny and the lasting contribution that it is likely to make to the future of Indian civilisation'. There was no good in looking back to an imaginary golden age, 'when nobody is supposed to have had anything to complain of', or of blaming science for all society's current ills. One of the solutions the backward-looking leaders offered was the 'total rejection of all modern technic for manufacture of the necessities of life, based upon the application of scientific knowledge'. It was true, the editorial conceded, that large numbers of Indians had suffered as a result of industrialisation, but that was not a reason to blame science and technology, only the manner in which they had been employed. Referring to Gandhi, the editorial continued, 'we do not for a moment subscribe [to the view] that better and happier conditions can be created by discarding modern scientific technic and reverting back to the spinning-wheel, the loin cloth and the bullock cart'. On the contrary, if the discoveries of science were 'properly and intensively applied' they would offer 'far better solutions to our bewildering economic, social and even political problems'.[103]

Indian scientists and their politician allies did not necessarily accept the gloomy prognosis offered by many colonial and metropolitan experts in the 1930s and 1940s, though they did take over the vision of an India burdened by poverty, hunger and disease. Although inclined to blame colonial self-interest or negligence for India's many disorders, they were disposed to believe that science and technology, rightly applied, could rapidly transform India's fortunes and remove many of its longstanding problems. Writing from his European exile in October 1935, Subhas Chandra Bose welcomed *Science and Culture* and the contribution scientific knowledge could make to tackling

---

[103] *Science and Culture*, 1 (1935), pp. 3–4.

India's needs. 'Whatever might have been the views of our older "Nation-builders"', he remarked, no doubt with Gandhi in mind, 'we younger folks approach the task of nation-building in a thoroughly scientific spirit and we desire to be armed with all the knowledge which modern science and culture can afford us.'[104] By the late 1930s the ideals and achievements of Roosevelt's New Deal and state planning in Soviet Russia had percolated through to India, nurturing a new faith in the potentialities of state-directed science and technology. One of the foremost advocates was Saha, who persuaded Bose and subsequently Nehru, the two leading figures among the younger generation of Indian nationalists, to take an active interest in the application of modern science and technology to India's national future. The Congress, holding ministerial power in eight provinces since the elections of 1937 had inaugurated the working of a new constitution under the Government of India Act of 1935, saw the need to reverse the effects of excessive provincialisation by forming central bodies to coordinate India-wide policy for vital areas such as public health.[105] In 1938 the Congress went a stage further, setting up (at Saha's suggestion) a National Planning Committee. Although it lost impetus once the Congress ministries resigned on the outbreak of the Second World War and ceased to function with Nehru's imprisonment in 1941, the committee marked the beginning of the party's active commitment to state planning and the use of science and technology to serve nation-building goals.

In his own speeches and writings in the late 1930s and 1940s Saha brought together many of the issues that had been uppermost in the minds of many Bengali scientists and intellectuals in recent decades: how to revive the ailing countryside of Bengal through village regeneration schemes, flood control and malaria eradication, and measures to increase both food production and industrial output. Saha argued in 1938 that though India had a population of 360 million it was not, as European commentators believed, overpopulated – so long as it could put its resources and expertise to work. India had coal, water power and other natural resources in abundance; it remained poor only 'because these resources have not been developed and industrial work has not been properly planned and organised'. If the proper measures were undertaken, India could at last 'lay the foundations of a strong and prosperous national life'.[106]

In September 1943, as famine engulfed Bengal, another leading member of

---

[104] Subhas Chandra Bose, 'Some Problems of Nation-Building', *Science and Culture*, 1 (1935), p. 258.
[105] Education, Health and Lands (Health), 63–1, 1938, NAI.
[106] Meghnad Saha, *Annual Address to the National Institute of Sciences of India* (Calcutta, 1938).

the *Science and Culture* group, J. C. Ghosh, spoke in similar terms at a meeting of the National Institute of Sciences. He denounced those 'powerful leaders' who seemed 'so impressed by the evils of the modern world' that they were opposed to all attempts to use state power to aid scientific and technological development. They apparently preferred 'the culture of cottage industries, peasant farming, and living on subsistence level with the inevitable doses of famine and pestilence'. It was, therefore, the duty of men of science in India to shoulder the additional burden of 'educating their masters out of this conservative view'. Science could be misused, but it could also be a force for 'immense good' if handled for the benefit of humanity. He quoted Pasteur: 'Science is the soul of all progress and the source of all prosperity.'[107]

In a speech in London the following year, Saha returned to the same themes, but with an even greater sense of urgency and opportunity in the wake of famine in Bengal and the Beveridge Report in Britain. Saha accepted the view that 90 per cent of India was 'still in the Middle Ages' and had only a 'thin veneer of modernism'. The National Planning Committee had shown that India was 'almost entirely untouched by modern scientific methods'. If the country genuinely wanted to advance it must tackle 'the great task of applying modern scientific and industrial methods for the development of her potential', as had been done by the USSR, 'with signal success', in the past twenty-five years. The Bengal famine had underscored the message that India had too little industry and relied too heavily on a primitive system of agriculture. He envisaged a project on the lines of the Tennessee Valley Authority for the Damodar valley in western Bengal and similar 'multi-purpose' developments elsewhere along India's major rivers. For the problems of India's millions to be solved, there had to be 'a national purpose behind all planning, and I do not see how any planning can be given without a National Government, or unless we have a Government which has popular support and is composed of leaders in whom the people have confidence'.[108]

Three years later India was partitioned and, in bloodshed and turmoil, the new states of India and Pakistan were born. In India, Nehru as Prime Minister lost no time in giving Indian science the authority and the opportunities it had craved for decades. On 15 August, within hours of India's 'freedom at midnight', a portfolio for scientific research was created under the Prime Minister, symbolising his deep attachment to science and technology as the key to India's

---

[107] J. C. Ghosh, *National Institute of Sciences: Opening Address* (Calcutta, 1943), pp. 1–2.
[108] Meghnad Saha, 'Science in Social and International Planning, with Special Reference to India', *Nature*, 24 February 1945, pp. 221–4.

modernity and his long association with the Indian Science Congress.[109] In June 1948 a Department of Scientific Research was established, expanded in January 1951 into the Ministry of National Resources and Scientific Research. The IMS was abolished, IRFA was redesignated the Indian Council of Medical Research, and in March 1950 state planning was launched with the creation of the Planning Commission, in which by 1955 both the statistician P. C. Mahalanobis and the chemist J. C. Ghosh had significant roles. Nehru also had close relations with S. S. Bhatnagar, as the first Director-General of the Council of Scientific and Industrial Research, and Homi Bhabha, as Chairman of India's Atomic Energy Commission. Under Nehru, it seemed, India's national science had at last come of age. Nehru himself celebrated the dawning of the new age when he declared, with an optimism that had long since drained from colonial rhetoric, that science alone gave India the means to solve 'the problems of hunger and poverty, of insanitation and illiteracy, of superstition and the deadening custom and tradition, of vast resources running to waste, of a rich country inhabited by starving people'. 'Who', he asked, 'could afford to ignore science today? At every turn, we have to seek its aid . . . The future belongs to science and to those who make friends with science.'[110]

[109] *Selected Works of Jawaharlal Nehru*, VIII (New Delhi, 1976), pp. 806–12; Ward Morehouse, 'Nehru and Science: The Vision of New India', *Indian Journal of Public Administration*, 15 (1969), pp. 489–508.
[110] Government of India, *Scientific Research* (Delhi, 1957), p. 1.

CONCLUSION

The history of science, technology and medicine in India during the colonial period, so often in the past treated as marginal, can be seen in the light of recent scholarship as having a far more central, but also far more complex, role. The more historians take into account the ideological dimensions of science, technology and medicine, the more we move from seeing them as 'tools of empire' to explore their social, cultural and political dimensions, the more apparent it becomes that there was no simple, one-directional process of scientific and technological 'transfer', but rather a series of cross-cultural exchanges and interactions.

If we take what has been categorised by Basalla and others as 'colonial science', we can see at work not only the extension to British India of metropolitan agencies and ideas but, no less powerfully, the strength of the British interaction with India's culture and environment, and the consequent distancing of colonial from metropolitan science. In part this derived from a sense of the provincialism that divided Calcutta from London, but it also reflected the persistent belief that science, technology and medicine in India could not be identical with metropolitan models but needed to reflect local conditions and circumstances and the political imperatives of the colonial regime itself. Science was about India as much as being in India, as demonstrated by the nineteenth-century emphasis upon natural science and the scientific reconnaissance of the Indian landscape. India was understood and represented through science and medicine as an alien territory inhabited by a foreign race: the monsoon, the Himalayas, India's tropical diseases, the 'peculiar geography of Hindoostan', the manner of Indian pilgrimages, diets, marriage customs and purdah – the physical and cultural idiosyncrasies of India constantly flickered across the imperial vision of India.

Repeated borrowings from, and interactions with, Indian science were also a factor in determining the content of Western science in India and informing its eclectic outlook. Indian agency was indispensable, from the collection of botanical specimens and materia medica to surveying or running hospitals. None the less, there remained a sense in which, in imperial eyes, science belonged uniquely to the ruling race. From the 1780s through to the 1920s, science among Europeans in India was more than a profession conducted at the 'periphery'. Despite the equivocations of the Company, science had an evident utility for agriculture, revenue, trade and health. For the civil servant, as for the scientific practitioner, science symbolised reason and order, the colonial power's ability to 'know' India and bring it under effective and productive control. But, no less significantly, science was also part of the self-identity of the European elite and its self-declared mission to 'improve', to 'civilise', ultimately to 'modernise', India. Paradoxically, such 'improving' strategies nurtured a corresponding denial, predicated on race and culture, of Indians' right and ability to practise 'real' science or to assume a position of equality and authority within the institutional and intellectual arenas of colonial science. Indian scientists often had a more positive reception in London, Edinburgh, Berlin or Stockholm than they received in their own country at the hands of the white scientific and medical establishment. In science, technology and medicine, denial was as important as dissemination; the colonial pursuit of modernity was ever tinged with a suspicion that tradition suited India better. It is significant, too, that, after a highpoint of confidence in the transforming power of imperial science, technology and medicine in the late nineteenth and early twentieth centuries, by the 1920s and 1930s, with the rise to international recognition of Indian scientists and the challenge of the Indian nationalist movement, British scientists showed increasing pessimism about the contribution science could make to India's future.

In the late nineteenth and early twentieth centuries, Indian scientists confronted in extreme form many of the dilemmas that had previously vexed India's European scientists, above all the problem of how to reconcile localism with cosmopolitanism. Many middle-class Indians, and not just those in the burgeoning scientific establishment, saw Western science as empowering, a passport to the modern world to which they were entitled and that conferred on them the right to be equal participants and beneficiaries. They saw, at the same time, the possibility of making modern science their own, and not merely a derivative practice, by turning away from the European pursuit of natural science and the external understanding of India it represented, developing chemistry, physics and mathematics instead and aligning science, technology and medicine with the needs of the nation rather than the imperial power. In this endeavour they attained a significant degree of international recognition, symbolised by Raman's receipt of the Nobel Prize for Physics.

However, in making science Indian and engaging positively with India's own traditions and philosophy, in medicine as in physics, Indian scientists ran the risk of becoming ostracised from the international scientific community. Moreover, although Indian science has often been seen as coming of age in the 1890s and 1900s, in reality it continued to face major challenges. Internally, it faced forces opposed to the identification of the Indian nation with modern science, technology and medicine and so challenged the hegemonic leadership of India's scientific community or fostered its own ambivalent stance towards modernity. No less significantly, as long as the colonial order remained intact, it was hard for India's scientists to attain a commanding position in institutions such as the Indian Medical Service or to have a decisive say in science policy. Metropolitan influences from Britain, and increasingly after 1918 from the United States, helped to maintain the institutionally dependent character of Indian science. In the face of colonial parsimony, racial discrimination and the pressures of provincialisation, Indian scientists had to build new agencies and organisations of their own and to look for alternative political patronage from the modernisers among India's nationalist leadership.

But, as with the colonialists, the significance of science, technology and medicine for Indians ranged far beyond institutional arenas and governmental sites. Science, technology and medicine found some of their leading patrons among Indian princes, Parsi philanthropists, and the landed, mercantile and industrial elites. They provided employment for Indians as they did for Europeans, for men, but also in lesser numbers for women. They helped fashion the lifestyles and world-view of the middle classes. In campaigns against the sanitary 'ignorance' of the masses, the 'barbarism' of *dais*, or the 'quackery' of folk healers, Indian elites devised their own reform agendas, assembled their own hegemonic credentials. Railways, irrigation canals, dispensaries and hospitals, the printing press, the cinema and the motor bus had a material and cultural influence that reached far beyond the Western-educated middle classes. Despite the resistance that at times confronted technological innovation, diets and birthing practices, ideas of gender and community, race and nation, all in various ways and by different routes came to feel the impact of scientific, technological and medical change. Science by 1947 had come to assume a public importance, a social impact and a cultural resonance inconceivable at the start of the colonial era.

BIOGRAPHICAL NOTES

Aykroyd, W. R., 1899–1979. b. Dublin; Trinity College, Dublin; MD, 1928. Health Section of League of Nations, 1931–5; Director, Nutrition Research Laboratories, Coonoor, 1935–46; member, Bengal Famine Commission, 1944–5; Director of Nutrition Division, FAO, 1946–60.

Bhatnagar, Shanti Swarup, 1895–1955. b. Shahpur district, Punjab; Forman Christian College, Lahore; B.Sc., 1916; D.Sc., University College London. Professor of Chemistry, Benares Hindu University, 1921–4; Professor of Chemistry, University of Punjab, Lahore, 1924–40; Director, Board of Scientific and Industrial Research, Delhi, 1940f. Knighted, 1941; FRS, 1943; President, Indian Science Congress, 1944.

Bose, Jagadis Chandra, 1858–1937. b. Vikrampur, near Dacca; St Xavier's, Calcutta; Cambridge; D.Sc., London. Indian Educational Service, 1885–1915; began research on electric waves, 1895; plant physiology, 1899; founded Bose Institute, 1917. Knighted, 1915; FRS, 1920; President, Indian Science Congress, 1927.

Bose, Pramatha Nath, 1855–1934. b. Gaipur village, Twenty-Four Parganas; St Xavier's College, Calcutta; Gilchrist scholarship, 1874; B.Sc., London. Joined Geological Survey of India, 1880, as assistant superintendent; retired, 1903, as deputy superintendent. Presided over Industrial Conference, Calcutta, 1891.

Brandis, Dietrich, 1824–1907. Studied at Copenhagen, Gottingen and Bonn (Ph.D. in botany). Superintendent of Forests, Pegu, 1856; Inspector-General of Indian Forests, 1864–83; knighted, 1887.

Buchanan, Francis, 1762–1829. b. Brenziet, near Stirling; Glasgow; MD, Edinburgh, 1783. Assistant surgeon, Bengal Medical Service, 1794; surveys of south India, 1800–1, Bengal, 1807–14; Superintendent, Calcutta Botanic Garden, 1814–15; left India, 1815. Took name Buchanan-Hamilton, 1818.

Cautley, Proby T., 1802–71. b. Stratford St Mary, Suffolk. Joined Bengal Artillery, 1819; worked on Doab Canal, 1820–30s; Superintendent, Ganges Canal, 1842–54. With Falconer awarded Wollaston Medal of the Geological Society of London, 1837; knighted, 1854; member, Council of India, London, 1858–66.

Chopra, Ram Nath, 1882–1973. b. in Jammu and Kashmir; MD, 1912. Joined Indian Medical Service; Professor of Pharmacology, Calcutta School of Tropical Medicine, and Professor of Pharmacology, Calcutta Medical College, 1921f; chaired Drugs Enquiry Commission, 1930; retired, 1941.

Everest, George, 1790–1866. b. Gwenvale, Brecknockshire; joined Bengal Artillery, 1806; captain, 1818; major, 1832; lieutenant-colonel, 1838. Chief assistant, Trigonometrical Survey, 1818–23; Superintendent, 1823–43; Surveyor-General of India, 1830–43; retired, 1843. FRS, 1827; knighted, 1861; member, Council of Royal Society, 1863–5.

Falconer, Hugh, 1808–65. b. Forres, Morayshire; MD, Edinburgh, 1829. Assistant surgeon, Bengal Medical Service, 1830; Superintendent, Saharanpur Botanic Garden,

1832–41; Superintendent, Calcutta Botanic Garden, and Professor of Botany, Calcutta, 1848–54; left India, 1855. With Cautley awarded Wollaston Medal of the Geological Society of London, 1837; FLS, 1844; FRS, 1845.

Hodgson, B. H., 1800–94. b. Prestbury, Cheshire; Haileybury College. Served in India, 1818f; Resident, Nepal, 1833–43; lived at Darjeeling, 1845–58; published extensively on Himalayan zoology and ethnology; left India, 1858. FLS, 1835; FRS, 1877.

Holland, Thomas Henry, 1868–1947. b. Helston, Cornwall; Royal School of Mines. Entered Geological Survey of India as assistant superintendent, 1890; Director, 1903–10; Professor of Geology, Manchester University, 1910–18; President, Indian Industrial Commission, 1916–18; member, Viceroy's Executive Council, 1920–21. Rector, Imperial College, London, 1922–29; Principal and Vice-Chancellor, Edinburgh University, 1929–44. FRS, 1904; knighted, 1908.

Hooker, Joseph Dalton, 1817–1911. b. Halesworth, Suffolk, son of Sir William Jackson Hooker; Glasgow University. Expeditions to Antarctica and New Zealand, 1840–43, and India, 1848–50. Succeeded his father as Director of Kew Gardens, 1841–85.

Howard, Albert, 1873–1947. b. Shropshire; Cambridge; MA, 1902. Lecturer in Agricultural Science, Harrison College, Barbados, 1899–1902; Imperial Economic Botanist to Government of India, 1905–24; Director, Indore Institute of Plant Industry, 1924–31. President, Indian Science Congress, 1926; knighted, 1934.

Lambton, William, 1756–1823. b. Crosby Grange, Yorkshire. Entered army, 1781; transferred to HM 33rd Foot, 1794; captain, 1806, major, 1806, lieutenant-colonel, 1814. Served in India, 1798f; survey of south India, 1800–18; Superintendent, Trigonometrical Survey of India, 1818–23. FRS.

McCarrison, Robert, 1878–1960. b. Portadown, Northern Ireland; Queen's, Belfast. Entered Indian Medical Service, 1901; served in Chitral and Gilgit, 1902–11; research on goitre. Began beri-beri enquiry, 1918; invalided to Britain, 1920–2; headed Deficiency Disease Enquiry, 1925–9; Director, Nutrition Research Laboratories, Coonoor, 1929–35. Knighted, 1933.

Mackenzie, Colin, 1754–1821. b. Isle of Lewis. Served with Madras Engineers, 1782f; captain, 1793, major, 1806, lieutenant-colonel, 1810, colonel, 1819. Survey of Nizam's Dominions, 1792, Mysore, 1799–1810; Surveyor-General, Madras, 1810, 1814–15; Surveyor-General of India, 1815–21.

Martin, James Ranald, 1796–1874. b. Isle of Skye; St Bartholomew's Hospital, London. Assistant surgeon, Bengal Medical Service, 1817; surgeon, 1828; Presidency Surgeon, 1830; left India, 1840. Member, Royal Commission on Sanitary State of the Army in India, 1859–63; knighted, 1860.

O'Shaughnessy, W. B., 1809–89. b. Limerick, Ireland; MD, Edinburgh, 1830. Assistant surgeon, Bengal Medical Service, 1833; surgeon, 1848; Professor of Chemistry, Calcutta Medical College; Director-General of Telegraphs, 1852f; left India, 1861. FRS, 1843; knighted, 1856.

Raman, Chandrasekhara Venkata, 1888–1970. b. Trichinopoly; Presidency College, Madras, MA (Mathematics), 1907. Joined Indian Finance Department; Palit Professor of Physics, Calcutta University, 1917–33; discovered 'Raman effect', 1928; Director of Indian Institute of Science, Bangalore, 1933–7; founded Raman Research Institute, 1948. FRS, 1924; knighted, 1929; Nobel Prize for Physics, 1930.

Ray, Prafulla Chandra, 1861–1944. b. Rarul village, Khulna district, Bengal; Presidency

College, Calcutta; Gilchrist scholarship, 1882; B.Sc., Edinburgh, 1885; D.Sc., 1887. Professor of Chemistry, Presidency College, Calcutta, 1889f; Palit Professor of Chemistry, University of Calcutta, 1917. Knighted, 1919; President, Indian Science Congress, 1920; President, Indian Chemical Society, 1924–6.

Ross, Ronald, 1857–1932. b. Almora, North-Western Provinces, son of General Sir Campbell Ross. MRCS; St Bartholomew's Hospital, London. Indian Medical Service, 1881–99; joined Liverpool School of Tropical Medicine, 1899, Professor, 1902–12. FRS, 1901; Nobel Prize for Medicine, 1902; knighted, 1911.

Roxburgh, William, 1751–1815. b. Craigie, Ayrshire; MD, Edinburgh. Surgeon's mate on East India Company ships, 1766f; joined Madras Medical Service as assistant surgeon, 1776; surgeon, 1780. Superintendent, Samalkot Botanic Garden, 1781–93; Superintendent, Calcutta Botanic Garden, 1793–1813; retired, 1813. FLS, 1799.

Royle, John Forbes, 1799–1858. b. Cawnpore; Edinburgh; MD, Munich, 1833. Assistant surgeon, Bengal Medical Service, 1819; Superintendent, Saharanpur Botanic Garden, 1823–31; retired to England, 1831. Professor of Materia Medica and Therapeutics, King's College, London, 1837–56. FLS, 1833; FRS, 1837; Secretary, Geological Society of London, 1836–8.

Saha, Meghnad, 1893–1956. b. Seoratali village near Dacca; B.Sc., Presidency College, Calcutta; M.Sc., Calcutta, 1915; D. Phil., Berlin. Professor of Physics, Calcutta, 1921; Professor of Physics, Allahabad, 1922f. Member, National Planning Committee, 1938f. FRS, 1927; President, Indian Science Congress, 1934.

Sircar, Mahendralal, 1833–1904. b. village near Howrah; Calcutta Medical College; MD, 1863. Converted to homoeopathy, 1867; launched *Calcutta Journal of Medicine*, 1868; founded Indian Association for the Cultivation of Science, 1876.

Wallich, Nathaniel, 1786–1854. b. Copenhagen; MD; joined Danish Medical Service. Appointed surgeon to Danish settlement of Serampore, 1807, but joined East India Company's medical service after its capture, 1813; Superintendent, Calcutta Botanic Garden, 1815f; left India, 1847. FLS, 1818; FRS, 1829.

BIBLIOGRAPHICAL ESSAY

1. General

There have been few general surveys of the history of science, technology and medicine in colonial India and a dearth of interpretative essays. Standard histories of science, technology and medicine written from the perspective of Europe and North America give little coverage to India. At most there might be some initial acknowledgement of the mathematics, medicine, chemistry and astronomy of ancient India, but Joseph Needham's work on China (unparalleled in range and quality for India) is more often cited by authors in search of non-European comparisons. Of the specifically Indian works, D. M. Bose, S. N. Sen and B. V. Subbarayappa (eds.), *A Concise History of Science in India* (New Delhi, 1971), provides a convenient overview of a longer period than that covered by this book, but the chapter by Subbarayappa 'Western Science in India up to the End of the Nineteenth Century AD' (pp. 484–567) is a useful summary across several scientific fields. Unfortunately, no attempt is made to cover the twentieth century or to discuss medicine and technology.

There has been a tendency to partition the study of India's science, technology and medicine, like much else in the region's history, along conventional lines into ancient (Hindu), medieval (Muslim) and modern (colonial) periods. Of works that do link the pre-colonial and colonial periods of Indian scientific history, particularly useful are Ahsan Jan Qaisar, *The Indian Response to European Technology and Culture (AD 1498–1707)* (Delhi, 1982); and two articles by S. N. Sen, 'Scientific Works in Sanskrit, Translated into Foreign Languages and Vice Versa in the 18th and 19th Century AD', *IJHS*, 7, 1972, pp. 44–70, and 'The Character of the Introduction of Western Science in India during the Eighteenth and Nineteenth Centuries', *IJHS*, 1, 1966, pp. 112–22. The journal, especially in its early years, contains a number of useful articles on this and related themes: worth also noting here is R. L. Verma, 'The Growth of Greco-Arabian Medicine in Medieval India', *IJHS*, 5, 1970, pp. 347–63. Several articles on technology by Irfan Habib are also important in providing a bridge between earlier and later history, especially 'The Technology and Economy of Mughal India', *IESHR*, 18, 1980, pp. 1–34, and 'Technological Change and Society, 13th and 14th Centuries', in Debiprasad Chattopadhyaya (ed.), *Studies in the History of Science in India*, II (New Delhi, 1982), pp. 816–44; this work reproduces a number of other useful articles, including a survey article by A. Rahman on 'Science and Technology in Medieval India' (pp. 805–15).

Returning to general works, the fifteen volumes of O. P. Jaggi's *History of Science, Technology and Medicine in India* (Delhi, 1969–84), especially volumes x–xv, which look at science, technology and medicine in the modern period, bring together a vast amount of material and provide useful bibliographical references, but offer little structured analysis or critical interpretation. G. Kuppuram and K. Kumudamani (eds.), *History of*

Science and Technology in India (Delhi, 1990), is another multi-volume work, reproducing a number of articles (of varying quality and utility) previously published elsewhere. Volume I on *Health and Medicine, Science and Religion* has some relevant material; volume v on *Science and Technology* contains substantial essays by A. Rahman, Satpal Sangwan and others. There remains, however, much useful and otherwise inaccessible material in several standard 'colonial' works, notably D. G. Crawford, *A History of the Indian Medical Service* (2 volumes, London, 1914); G. W. MacGeorge, *Ways and Works in India* (Westminster, 1894); E. W. C. Sandes, *The Military Engineer in India* (2 volumes, Chatham, 1935); and R. H. Phillimore, *Historical Records of the Survey of India* (4 volumes, Dehra Dun, 1945–58). Of more recent works, Deepak Kumar, *Science and the Raj, 1857–1905* (Delhi, 1995), offers the best single introduction to colonial science policy and research. It draws on much original research and covers more of the Company period than the title suggests. Zaheer Baber's *The Science of Empire: Scientific Knowledge, Civilisation, and Colonial Rule in India* (Delhi, 1998) attempts a general overview at a more abstract level of discussion but without the benefit of substantial new research. It does, however, provide a useful introduction to the pre-colonial period. There is also useful material in two recent collections edited by Deepak Kumar: *Science and Empire: Essays in Indian Context* (Delhi, 1991) and (with Roy MacLeod) *Technology and the Raj: Western Technology and Technical Transfers to India, 1700–1947* (New Delhi, 1995), both of which will be referred to again in later sections of this bibliographical essay.

2. Science under the Company

Once a largely neglected field for the history of science, the period of Company rule in India has of late produced a number of interesting and important studies that have begun substantially to qualify and recontextualise much earlier work. Thus, in the field of botany and environmental science, I. H. Burkhill's *Chapters on the History of Botany in India* (Delhi, 1965) now needs to be considered alongside such works as Ray Desmond, *The European Discovery of the Indian Flora* (Oxford, 1992); Richard H. Grove, *Green Imperialism: Colonial Expansion, Tropical Island Edens and the Origins of Environmentalism, 1600–1860* (Cambridge, 1995) and his article 'Conserving Eden: The (European) East India Companies and their Environmental Policies on St. Helena, Mauritius and in Western India, 1660–1854', *Comparative Studies in Society and History*, 35, 1993, pp. 318–51; David Mackay, *In the Wake of Cook: Exploration, Science and Empire, 1780–1801* (London, 1985); and Lucile H. Brockway, *Science and Colonial Expansion: The Rise of the British Royal Botanic Gardens* (New York, 1979). Environmental history and the large volume of work on forestry, which has concentrated more on the late rather than early nineteenth century, will be noted in section 5 of this bibliography. In the field of geology, notable new work includes Andrew Grout, 'Geology and India, 1775–1805: An Episode in Colonial Science', *SAR*, 10, 1990, pp. 1–18, and his 1995 SOAS Ph.D., 'Geology and India, 1770–1851: A Study in the Methods and Motivations of a Colonial Science', as well as Satpal Sangwan, 'Reordering the Earth: The Emergence of Geology as a Scientific Discipline in Colonial India', *IESHR*, 31, 1994, pp. 291–310. The work of the Trigonometrical Survey has been critically re-examined by Matthew H. Edney in *Mapping an Empire: The Geographical Construction of British India, 1765–1843* (Chicago, 1997), a work that, with Grout's thesis, may serve as a model for other in-depth studies of specific branches of Company and colonial science in India. C. A. Bayly's *Empire and Information: Intelligence Gathering and Social Communication in India, 1780–1870* (Cambridge, 1996), although not cast in the conventional history of science mould, offers an important reinterpretation of Orientalist science and the early interaction between European and Indian science and medicine.

By contrast with writing about the history of science in Europe and North America, relatively little work has yet been published on either scientific institutions or individual scientists. Marika Vicziany's essay on Buchanan ('Imperialism, Botany and Statistics in Early Nineteenth-Century India: The Surveys of Francis Buchanan, 1762–1829', *MAS*, 20, 1986, pp. 625–60) is a significant exception, but does not entirely supersede David Prain's memoir, 'A Sketch of the Life of Francis Hamilton' in *Annals of the Royal Botanic Gardens, Calcutta*, 10, 1905, pp. i–lxxv, and needs to be read alongside Peter Robb, 'Completing "Our Stock of Geography", Or an Object "Still More Sublime": Colin Mackenzie's Survey of Mysore, 1790–1810', *JRAS*, 8, 1998, pp. 181–206. There remains much valuable material in older biographical works, notably William Wilson Hunter's *Life of Brian Houghton Hodgson* (London, 1896). More detailed studies of figures such as Roxborough, Wallich and Royle would be invaluable, but see Satpal Sangwan, 'The Strength of Scientific Culture: Interpreting Disorder in Colonial Science', *IESHR*, 34, 1997, pp. 217–50, for some of the issues and personalities involved. Useful, too, would be studies of the scientific contribution of the Asiatic Society of Calcutta and its counterparts in Bombay and Madras, whose journals remain a remarkably under-utilised source for the scientific ideas and practices of the nineteenth century. Grout's thesis offers some illumination on the Calcutta society, but, after more than a century, P. N. Bose's survey of 'Natural Science' in the *Centenary Review of the Asiatic Society of Bengal from 1784 to 1883* (Calcutta, 1885) remains a useful introduction to its activities. There is also much interesting material on science in India to be gleaned from the correspondence of two leading British scientists of the period: Leonard Huxley, *Life and Letters of Sir Joseph Dalton Hooker* (2 volumes, London, 1918), and Frederick Burkhardt and Sydney Smith (eds.), *The Correspondence of Charles Darwin*, volumes II–VII (Cambridge, 1986–91), covering the period 1837 to 1859, and subsequent volumes when they appear. Of the older secondary literature, there is still some value in H. J. C. Larwood, 'Western Science in India before 1850', *Journal of the Royal Asiatic Society of Great Britain and Ireland*, Part 2, 1962, pp. 62–76, and Mel Gorman, 'Introduction of Western Science into Colonial India: Role of the Calcutta Medical College', *Proceedings of the American Philosophical Society*, 132, 1988, pp. 276–98.

3. Colonial Medicine

D. G. Crawford's two-volume *History of the Indian Medical Service* (1914) remains an important source for the medical profession in colonial India, and, for biographical information, can be supplemented by his *Roll of the Indian Medical Service, 1615–1930* (Calcutta, 1930). Of continuing value, too, is Margaret I. Balfour and Ruth Young, *The Work of Medical Women in India* (London, 1929). But approaches to the subject of disease and medicine for nineteenth-century India have been substantially revised by a spate of recent works, notably Radhika Ramasubban, 'Imperial Health in British India, 1857–1900', in Roy MacLeod and Milton Lewis (eds.), *Disease, Medicine, and Empire:*

Perspectives on Western Medicine and the Experience of European Expansion (London, 1988), pp. 38–60; Roger Jeffery, *The Politics of Health in India* (Berkeley, CA, 1988), part 1 of which covers the colonial period; David Arnold, *Colonizing the Body: State Medicine and Epidemic Disease in Nineteenth-Century India* (Berkeley, CA, 1996); and Mark Harrison, *Public Health in British India: Anglo-Indian Preventive Medicine, 1859–1914* (Cambridge, 1994). Harrison's book and Anil Kumar, *Medicine and the Raj: British Medical Policy in India, 1835–1911* (New Delhi, 1998), also provide useful information about medical education and training.

On the subject of epidemic disease in the nineteenth and early twentieth centuries there has been a string of interesting and informative articles from Ira Klein, including 'Malaria and Mortality in Bengal', *IESHR*, 9, 1972, pp. 132–60; 'Death in India, 1871–1921', *JAS*, 22, 1973, pp. 639–59; and 'Cholera: Theory and Treatment in Nineteenth Century India', *Journal of Indian History*, 58, 1980, pp. 35–51. Cholera also forms the subject of a further essay by Mark Harrison, 'A Question of Locality: The Identity of Cholera in British India, 1860–1890', in David Arnold (ed.), *Warm Climates and Western Medicine: The Emergence of Tropical Medicine, 1500–1900* (Amsterdam, 1996), pp. 133–59, and a chapter in Sheldon Watts, *Epidemics in History: Disease, Power and Imperialism* (New Haven, CT, 1997), pp. 167–212. For the impact of another important epidemic, bubonic plague, see section 5 below.

On the medical front, there has been extensive discussion of the nineteenth-century interaction between Western and Indian medicine, notably Brahmananda Gupta, 'Indigenous Medicine in Nineteenth- and Twentieth-Century Bengal', in Charles Leslie (ed.), *Asian Medical Systems: A Comparative Study* (Berkeley, CA, 1976), pp. 368–78; John C. Hume, 'Rival Traditions: Western Medicine and Yunan-i Tibb in the Punjab, 1849–1889', *BHM*, 61, 1977, pp. 214–31; and Poonam Bala, *Imperialism and Medicine in Bengal: A Socio-Historical Perspective* (New Delhi, 1991). In addition to Balfour and Young's 1929 study, cited above, there has been rapidly growing interest in the gender dimensions of health and medicine in nineteenth-century India, as in Maneesha Lal, 'The Politics of Gender and Medicine in Colonial India: The Countess of Dufferin's Fund', *BHM*, 68, 1994, pp. 29–66, and Geraldine Forbes, 'Managing Midwifery in India', in Dagmar Engels and Shula Marks (eds.), *Contesting Colonial Hegemony: State and Society in Africa and India* (London, 1994), pp. 152–72. Forbes also remarks on the role of Indian women doctors in her volume for the *New Cambridge History of India: Women in Modern India* (Cambridge, 1996).

Discussion of the relationship between women and medicine in the colonial period has also focused on the contagious diseases Acts. The legislation and its rationale and impact have been the subject of several articles, among them Philippa Levine, 'Venereal Disease, Prostitution, and the Politics of Empire: The Case of British India', *Journal of the History of Sexuality*, 4, 1993–4, pp. 579–602, and Judy Whitehead, 'Bodies Clean and Unclean: Prostitution, Sanitary Legislation, and Respectable Femininity in Colonial North India', *Gender History*, 7, 1995, pp. 41–63. Douglas M. Peers, 'Soldiers, Surgeons and the Campaigns to Combat Sexually Transmitted Diseases in Colonial India, 1805–1860', *MH*, 42, 1998, pp. 137–60, reviews some of the earlier literature on the subject while adding to its military dimension. The health of the army, itself a major subject, is alluded to by many of the writers cited in this section (including Ramasubban's 'Imperial Health in British India' and Arnold's *Colonizing the Body*), but is examined most methodically by Philip D. Curtin, *Death by Migration: Europe's Encounter with the Tropical World in the Nineteenth Century* (Cambridge, 1989). On the emergence of sanitary policy and public health, there are insightful essays by J. B. Harrison, 'Allahabad: A Sanitary History', in K. Ballhatchet and J. B. Harrison (eds.), *The City in South Asia* (London, 1980), pp. 167–96; John C. Hume, 'Colonialism and Sanitary Medicine: The Development of Preventive Health Policy in the Punjab, 1860–1900', *MAS*, 20, 1986, pp. 703–24; and Mark Harrison, 'Towards a Sanitary Utopia? Professional Visions and Public Health in India, 1880–1914', *SAR*, 10, 1990, pp. 19–40.

4. Technology

Arnold Pacey's *Technology in World Civilization: A Thousand-Year History* (Oxford, 1990) places India in a wider comparative framework and raises a number of general issues, but draws on a rather restricted range of Indian examples. Daniel R. Headrick's two books, *Tools of Empire: Technology and European Imperialism in the Nineteenth Century* (New York, 1981) and the more substantial *The Tentacles of Progress: Technology Transfer in the Age of Imperialism, 1850–1940* (New York, 1988), offer a more provocative, but heavily Eurocentric, view of nineteenth- and early twentieth-century technology; *Tentacles of Progress* makes extensive use of Indian material, especially for railways and irrigation. By contrast, Dharampal's selection of European views, *Indian Science and Technology in the Eighteenth Century: Some Contemporary European Accounts* (Delhi, 1971), raises a number of issues about methodology and context left largely unanswered by his introductory essay. Michael Adas in *Machines as the Measure of Men: Science, Technology, and Ideologies of Western Dominance* (Ithaca, NY, 1989) more securely places Western views in context and makes useful comparisons with China and Africa, but, like Dharampal, his exclusive focus on Western sources allows little opportunity to evaluate Indian responses, one of the most persistent and regrettable lacunae in scholarly work on technology in colonial India.

There exist a number of studies on specific aspects of nineteenth-century technology. On shipping, the sub-title of Henry T. Bernstein's *Steamboats on the Ganges: An Exploration in the History of India's Modernization through Science and Technology* (Bombay, 1960) makes clear the author's technological presumptions from the outset. The essay by Sangwan Satpal, 'The Sinking Ships: Colonial Policy and the Decline of Indian Shipping, 1735–1835', in MacLeod and Kumar (eds.), *Technology and the Raj*, pp. 137–52, is one of several useful essays in that volume and a partial corrective to Bernstein. Satpal develops this and a number of related ideas about technology in his *Science, Technology and Colonisation: An Indian Experience, 1757–1857* (Delhi, 1991). Textile history has received attention in a number of studies, though most focus on the economic and organisational aspects of production: Hameeda Hossain, *The Company Weavers of Bengal: The East India Company and the Organization of Textile Production in Bengal, 1750–1813* (Delhi, 1988), is something of an exception in its coverage of textile technology. *Cloth and Commerce: Textiles in Colonial India* (Delhi, 1996), a collection of essays edited by Tirthankar Roy, also gives some account of production techniques. Discussion of silk production has become bogged down in the arcane issue of filatures, but Sabyasachi Bhattacharyya's 'Cultural and Social Constraints on Technological Innovation: Some Case Studies', *IESHR*, 3, 1966, pp. 240–67, remains, deservedly, a classic. Mining and metallurgy have received scant attention, and Valentine Ball's *A Manual of the Geology of India, Part III: Economic Geology* (Calcutta, 1881) still contains valuable information and a guide to the extensive journal literature not to be found elsewhere; some articles on the subject have appeared in *IJHS*.

Irrigation, by contrast, has begun to receive the attention it deserves, though for the broader, more technologically informed picture, apart from Headrick's *Tentacles of Progress*, it is still necessary to seek out older works such as Alfred Deakin's *Irrigated India: An Australian View of India and Ceylon* (London, 1893), and volume II of Sandes's *Military Engineer in India* (1935). Of the recent essays on irrigation, David Ludden, 'Patronage and Irrigation in Tamil Nadu: A Long-Term View', *IESHR*, 16, 1979, pp. 347–65, Nirmal Sengupta, 'The Indigenous Irrigation Organization of South Bihar', *IESHR*, 18, 1980, pp. 157–89, and David Hardiman, 'Well Irrigation in Gujarat: Systems of Use, Hierarchies of Control', *Economic and Political Weekly*, 20 June 1998, pp. 1533–44, are exemplary local studies. Ian Stone, *Canal Irrigation in British India: Perspectives on Technology Change in a Peasant Economy* (Cambridge, 1984), comes closest to an in-depth modern scholarly assessment of the technological importance and economic impact of colonial canal irrigation, but it needs to be read in conjunction with Elizabeth Whitcombe's essay on irrigation in Dharma Kumar (ed.), *The Cambridge Economic History of India*, volume II (Cambridge, 1983), pp. 677–737, and her account of 'The Environmental Costs of Irrigation in British India: Waterlogging, Salinity, Malaria', in David Arnold and Ramachandra Guha (eds.), *Nature, Culture, Imperialism: Essays on the Environmental History of South Asia* (Delhi, 1995), pp. 237–59, and R. J. Henry, 'Technology Transfer and Its Constraints: Early Warnings from Agricultural Development in Colonial India', in the MacLeod and Kumar volume, *Technology and the Raj*, pp. 51–77.

Railways, predictably, have excited the greatest interest of all nineteenth-century technologies, and of recent studies with a strong technological input Ian J. Kerr's *Building the Railways of the Raj, 1850–1900* (Delhi, 1995) is the most outstanding, but can still be supplemented by MacGeorge's *Ways and Works*, pp. 65–104. Also valuable are Fritz Lehmann's two essays, 'Great Britain and the Supply of Railway Locomotives to India: A Case Study of "Economic Imperialism"', *IESHR*, 2, 1965, pp. 297–306, and 'Railway Workshops, Technology Transfer, and Skilled Labour Recruitment in Colonial India', *Journal of Historical Research* (Ranchi), 20, 1977, pp. 49–61, but these are partly contradicted by Clive Dewey, 'Some Consequences of Military Expenditure in British India: The Case of the Upper Sagar Doab, 1849–1947', in Clive Dewey (ed.), *Arrested Development in India: The Historical Dimension* (Riverdale, MD, 1988), pp. 93–169, and Ian Derbyshire's essay 'The Building of India's Railways: The Application of Western Technology in the Colonial Periphery, 1850–1920', in MacLeod and Kumar's *Technology and the Raj*, pp. 177–215. This volume also includes Saroj Ghose, 'Commercial Needs and Military Necessities: The Telegraph in India', pp. 153–76, a subject also discussed in Headrick's *Tentacles of Progress*, chapter 4.

The survival of 'traditional' technologies in nineteenth-century India and the technological aspects of the emerging industries in India have yet to be adequately discussed, though they receive a barely nostalgic mention in such established works as D. R. Gadgil, *The Industrial Evolution of India in Recent Times, 1860–1939* (5th edition, Delhi, 1972). Nor has the debate about India's technological and industrial future by George Birdwood, E. B. Havell and Alfred Chatterton received much scholarly comment, though an important exception, from the Indian perspective, is Dhruv Raina and S. Irfan Habib, 'The Unfolding of an Engagement: *The Dawn* on Science, Technical Education and Industrialization: India, 1896–1912', *SH*, 9, 1993, pp. 87–117.

5. Science and Empire

Deepak Kumar's *Science and the Raj* provides the best account of the scientific policy and practice of the period, but his earlier article, 'Racial Discrimination and Science in Nineteenth Century India', *IESHR*, 19, 1982, pp. 63–82, remains an important, as yet unchallenged, statement. Valuable, too, from a more metropolitan perspective is Roy MacLeod, 'Scientific Advice for British India: Imperial Perceptions and Administrative Goals, 1898–1923', *MAS*, 9, 1975, pp. 343–84. The plague crisis of the 1890s and 1900s, which marked such an important turning point in colonial science, has been discussed by several authors including I. J. Catanach, 'Plague and the Tensions of Empire, India, 1896–1918', in David Arnold (ed.), *Imperial Medicine and Indigenous Societies* (Manchester, 1988), pp. 149–71; Ira Klein, 'Plague, Policy and Popular Unrest in British India', *MAS*, 22, 1988, pp. 723–55; Arnold, *Colonizing the Body*, chapter 5; and Rajnarayan Chandavarkar, 'Plague Panic and Epidemic Politics in India, 1896–1914', in Terence Ranger and Paul Slack (eds.), *Epidemics and Ideas: Essays on the Historical Perceptions of Pestilence* (Cambridge, 1992), pp. 203–40. Other aspects of the scientific history of the period (including botany, geology and agricultural and veterinary science) have, however, received scant attention, and the most helpful sources remain the colonial texts of the period, among the most significant being the *Indian Industrial Commission, 1916–18* (Calcutta, 1918), and the *Royal Commission on Agriculture in India* (London, 1928). As a guide to the former, Clive Dewey's 'The Government of India's "New Industrial Policy", 1900–1925: Formation and Failure', in K. N. Chaudhuri and Clive J. Dewey (eds.), *Economy and Society: Essays in Indian Economic and Social History* (Delhi, 1979), pp. 215–57, is invaluable.

One area where there has been a remarkable surge of new scholarship is in the field of environmental history and specifically forestry. Pioneering studies of note include Madhav Gadgil and Ramachandra Guha, 'State Forestry and Social Conflict in British India', first published in *Past and Present*, no. 123, 1989, pp. 141–77, and reproduced in David Hardiman (ed.), *Peasant Resistance in India, 1858–1914* (Delhi, 1992), pp. 259–95; the same authors' *This Fissured Land: An Ecological History of India* (Delhi, 1992); Richard Grove's *Green Imperialism*; and Mahesh Rangarajan, *Fencing the Forest: Conservation and Ecological Change in India's Central Provinces, 1860–1914* (Delhi, 1996). Two journals have been especially important in encouraging the growth of India's environmental history. Articles in the *Indian Economic and Social History Review* include R. P. Tucker, 'Forest Management and Imperial Politics: Thana District, Bombay, 1823–1887', *IESHR*, 16, 1979, pp. 273–300; Ramachandra Guha, 'An Early Environmental Debate: The Making of the 1878 Forest Act', *IESHR*, 27, 1990, pp. 65–84; and Mahesh Rangarajan, 'Imperial Agendas and India's Forests: The Early History of Indian Forestry, 1800–1878', *IESHR*, 31, 1994, pp. 147–67. The journal *Environment and History* has begun to publish important contributions to this field. Volume 1, 1995, has essays by Vinita Damodaran, 'Famine in a Forest Tract: Ecological Change and the Causes of the 1897 Famine in Chotanagpur, Northern India', pp. 129–58, and Michael Mann,

'Ecological Change in North India: Deforestation and Agrarian Distress in the Ganga-Jamna Doab, 1800–1850', pp. 201–20. A special issue on South Asia, volume 2, no. 2, 1996, includes Mahesh Rangarajan, 'Environmental Histories of South Asia: A Review Essay', pp. 129–43; K. Sivaramakrishnan, 'The Politics of Fire and Forest Regeneration in Colonial Bengal', pp. 145–94; and Indra Munshi Saldanha, 'Colonialism and Professionalism: A German Forester in India', pp. 195–219, which looks at the career of Dietrich Brandis. Two important collections of essays are David Arnold and Ramachandra Guha (eds.), *Nature, Culture, Imperialism* (1995), and Richard Grove, Vinita Damodaran and Satpal Sangwan (eds.), *Nature and the Orient: The Environmental History of South and South East Asia* (Delhi, 1998).

The emergence of the Indian scientific community in late nineteenth and early twentieth centuries has been surprisingly neglected, apart from the hagiographies. But Ashis Nandy's essay, 'Defiance and Conformity in Science: The World of Jagadis Chandra Bose', in his *Alternative Sciences: Creativity and Authenticity in Two Indian Scientists* (2nd edition, Delhi, 1995), pp. 17–87, supplements Patrick Geddes, *An Indian Pioneer of Science: The Life and Work of Sir Jagadis C. Bose* (London, 1920). There is a brief essay by Chittabrata Palit, on 'Mahendra Lal Sircar, 1833–1904', in Deepak Kumar (ed.), *Science and Empire*, pp. 152–60; and Kumar himself gives a general commentary on several leading Indian scientists in 'The "Culture" of Science and Colonial Culture, India, 1820–1920', *BJHS*, 29, 1996, pp. 195–209. S. Irfan Habib and Dhruv Raina, 'Copernicus, Columbus, Colonialism and the Role of Science in Nineteenth Century India', *Social Scientist*, no. 17, 1989, pp. 50–66, and V. V. Krishna, 'The Colonial "Model" and the Emergence of National Science in India, 1876–1920', in P. Petitjean et al. (eds.), *Science and Empire* (Dordrecht, 1992), pp. 57–72, both offer critiques of the Basalla approach and an interpretative overview of the Indian science of the period.

6. Hindu Science

The reconstitution and revival of Hindu science and its relationship with Western medicine have begun to attract some significant scholarly comment, including Gyan Prakash, 'Science between the Lines', in Shahid Amin and Dipesh Chakrabarty (eds.), *Subaltern Studies IX* (Delhi, 1996), pp. 59–82; Partha Chatterjee, 'The Disciplines in Colonial Bengal', in Partha Chatterjee (ed.), *Texts of Power: Emerging Disciplines in Colonial Bengal* (Minneapolis, MN, 1995), pp. 1–29. David Gosling, *Science and Religion in India* (Madras, 1976), also provides interesting background material, as does Tapan Raychaudhuri, *Europe Reconsidered: Perceptions of the West in Nineteenth Century Bengal* (Delhi, 1988).

Attention has, however, mainly focused on the revival of Ayurvedic medicine. For this there are a number of interesting pieces, including: chapter 2 of Poonam Bala's *Imperialism and Medicine in Bengal*; Brahmananda Gupta's 'Indigenous Medicine in Nineteenth- and Twentieth-Century Bengal', in Charles Leslie (ed.), *Asian Medical Systems* (pp. 368–78); and Paul R. Brass, 'The Politics of Ayurvedic Education: A Case Study of Revivalism and Modernization in India', in S. H. Rudolph and L. I. Rudolph (eds.), *Education and Politics in India* (Cambridge, MA, 1972), pp. 342–71. Charles Leslie's own important contributions include an essay on 'The Ambiguities of Medical Revivalism in Modern India', in the *Asian Medical Systems* volume, pp. 356–67; 'The

Modernization of Asian Medical Systems', in John J. Poggie and Robert N. Lynch (eds.), *Rethinking Modernization: Anthropological Perspectives* (Westport, CT, 1974), pp. 69–108; and 'Interpretations of Illness: Syncretism in Modern Ayurveda', in Charles Leslie and Allan Young (eds.), *Paths to Asian Medical Knowledge* (Berkeley, CA, 1992), pp. 177–208. There is also an interesting essay by Ashis Nandy and Shiv Visvanathan on 'Modern Medicine and Its Non-Modern Critics', first published in Steven Marglin and Frédérique Apffel-Marglin (eds.), *Dominating Knowledge* (Oxford, 1990), pp. 145–84, and reprinted in Visvanathan's *A Carnival for Science: Essays on Science, Technology and Development* (Delhi, 1997), pp. 94–145. Regional studies of Ayurveda are increasingly important in assessing complexity and diversity within 'indigenous medicine'. In addition to the just cited works dealing with Bengal, there are two particularly informative studies for Kerala: Kizhedath Vasudevan Nair, *A Biography of Vaidyaratnam P. S. Varier* (Kottakkal, 1954); and K. N. Panikkar, 'Indigenous Medicine and Cultural Hegemony: A Study of the Revitalization Movement in Keralam', *SH*, 8, 1992, pp. 283–308. Other medical ideas and practices have been less extensively discussed, but on homoeopathy see Surinder M. Bhardwaj, 'Homoeopathy in India', in Giri Raj Gupta (ed.), *The Social and Cultural Context of Medicine in India* (Delhi, 1981), pp. 31–54; and for a leading practitioner and exponent of Unani medicine, Barbara Metcalf, 'Nationalist Muslims in British India: The Case of Hakim Ajmal Khan', *MAS*, 19, 1985, pp. 1–28.

7. Science, State and Nation

Despite their obvious importance, the science, technology and medicine of the period between the outbreak of the First World War and Indian independence have, as yet, attracted little critical scholarship. For a general summary of the period up to the outbreak of the Second World War, B. Prashad (ed.), *The Progress of Science in India during the Past Twenty-Five Years* (Calcutta, 1938), remains immensely useful both for the essays themselves and for the detailed bibliographical references provided. By contrast, C. G. Pandit and K. Someswara Rao, *Indian Research Fund Association and Indian Council of Medical Research, 1911–1961: Fifty Years of Progress* (Delhi, 1961), is celebratory and shallow. Official publications and reports, such as the *Royal Commission on Agriculture in India* (1928), the *Health and Development Committee* (4 volumes, 1946), and A. V. Hill's semi-official report on *Scientific Research in India* (1944), which can usefully be supplemented by his *The Ethical Dilemma of Science and Other Writings* (New York, 1960), remain invaluable. On scientists, there is an excellent study of two leading figures, Meghnad Saha and Homi Bhabha, in Robert S. Anderson, *Building Scientific Institutions in India: Saha and Bhabha* (Montreal, 1975), which also does much to illuminate the scientific rivalries of the inter-war years. Norah Richards' *Sir Shanti Swarup Bhatnagar: A Biographical Study of India's Eminent Scientist* (New Delhi, 1948) is dated but still useful, while the several volumes of the *Biographical Memoirs of Fellows of the Indian National Science Academy* (New Delhi, 1976) contain a number of potted scientific biographies not otherwise easily accessible. Ashis Nandy's account of the pioneer psychoanalyst Giridrasekhar Bose in *The Savage Freud and Other Essays on Possible and Retrievable Selves* (Princeton, NT, 1995), pp. 81–144, is disappointing for anyone in search of insights into the Indian scientific community of the time. Medical institutions have also been surprisingly neglected. For one significant contribution, see Helen Power, 'The Calcutta School of Tropical

Medicine: Institutionalizing Medical Research in the Periphery', *MH*, 40, 1996, pp. 197–214.

Relations between science, the colonial state and the nationalist movement are touched on in Jagdish N. Sinha, 'Science and the Indian National Congress', in Deepak Kumar (ed.), *Science and Empire*, pp. 161–81, and in his article 'Technology for National Reconstruction: The National Planning Committee, 1938–49', in Macleod and Kumar (eds.), *Technology and the Raj*, pp. 250–64. This volume also contains essays by Dinesh Abrol, '"Colonised Minds" or Progressive Nationalist Scientists: The Science and Culture Group', pp. 265–88, and, on the Council of Scientific and Industrial Research, V. V. Krishna, 'Organisation of Industrial Research; The Early History of CSIR, 1934–47', pp. 289–323. For medicine see Roger Jeffery, 'Recognizing India's Doctors: The Establishment of Medical Dependency, 1918–39', *MAS*, 13, 1979, pp. 245–70, and his 'Doctors and Congress: The Role of Medical Men and Medical Politics in Indian Nationalism', in Mike Shepperdson and Colin Simmons (eds.), *The Indian National Congress and the Political Economy of India, 1885–1985* (Aldershot, 1988), pp. 160–73. The changes in power and transportation and in industrial and agricultural technology in the period after 1914 receive some consideration in Amiya Kumar Bagchi, *Private Investment in India, 1900–1939* (Cambridge, 1972), and Rajat K. Ray, *Industrialization in India: Growth and Conflict in the Private Corporate Sector, 1914–47* (Delhi, 1979) and in regional studies such as Christopher John Baker, *An Indian Rural Economy, 1880–1955: The Tamilnad Countryside* (Oxford, 1984). In most of these studies, apart from Clive Dewey's article on Punjab noted earlier ('Some Consequences of Military Expenditure'), the impact of the Second World War and its immediate aftermath remains remarkably unexplored.

INDEX

Adas, Michael, 11
Age of Consent Bill, 91
Agra, 99, 113
Ajmer, 101
Andhra, 116, 178; University of, 191, 192
Agricultural Journal of India, 152
Agricultural and Horticultural Society of Calcutta, 29, 33, 49, 52, 150
agriculture, 5, 22, 23, 33, 35, 47, 51–2, 121, 125, 132–3, 137, 138, 150–3, 157, 163–4, 186, 187, 195, 203
Ainslie, Whitelaw, 67, 68
Akbar, 5, 36, 101
Ali, Syed Imdad, 157
Aligarh Scientific Society, 157
Aligarh, University of, 191
All-India Institute of Hygiene and Public Health, 194, 195, 200
Allahabad, 7, 59, 104, 106, 115, 206; University of, 191, 192
allopathy, 178–82, 184–5
American Methodist Episcopal Mission, 87
Amritsar, 7, 120
Andhra University, 191, 192
Annamalai University, 191
Annesley, James, 77, 78
anthropology, 155, 192
armaments, 5, 6–7, 100–1
Arms Act, 100
army engineers, 25, 26, 27, 116, 117–19, 156
Army Medical Department, 59
Arnold, William, 105
Arya Samaj, 170, 177
Arya Vaidya Samajam, 179
Asiatic Society of Bengal, 27, 28–9, 30–1, 32, 33, 34–5, 36, 37, 97, 155–7, 158, 159, 161, 162, 171, 172
Assam, 48, 51, 156
Association for the Advancement of Scientific and Industrial Education, 159
astronomy, 3, 5, 10, 35, 36–8, 41, 49, 134, 171
Atomic Energy Commission, 210
Australia, 11, 118
Awadh, 6, 8, 55, 109; Begum of, 122; Nawab of, 37
Aykroyd, W. R., 202, 214

Ayurveda, 4, 7, 62, 66–71, 172, 173, 175; revival of, 176–80, 182–3, 184, 192, 204
Ayurveda and Unani Tibbi College, 180

Bacteriological Department, 138, 144
bacteriology, 141, 142
Bahadur Shah, 124
Baker, W. E., 26, 27
Balfour, Margaret, 199
Ball, Valentine, 98, 139
Bangalore, 8, 161, 195, 205
Banjaras, 106, 107–8, 110
Banks, Sir Joseph, 20, 47, 52, 55, 97
Baroda, 26, 160
Basalla, George, 9–15, 16, 211
Basel Mission, 126
Basu, Baman Das, 65, 175
Bayly, C. A., 34, 36
Bellew, H. W., 87
Benares, 6, 7, 36, 106, 107, 115
Benares Hindu University, 187, 192
Bengal, 8, 22, 29, 31, 40, 45, 52, 53, 58, 68, 152, 158, 171, 208; disease in, 69, 71, 72, 78–80, 81; famine in, 46, 51, 78, 202, 203, 208, 209; Government of, 134, 167, 184; Partition of, 127, 159; technology in, 94–6, 104, 123, 126
Bengal Chemical and Pharmaceutical Works, 172
Bengal Engineers, 26, 119
Bengal Medical Service, 58
Bengal National College, 160
Bengali, scientific works in, 33, 38, 157–8, 168
Bentinck, Lord, 32, 63, 104
Berar, 41
Berthelot, Marcellin, 166
Beveridge Report, 204, 209
Bhabha, Homi, 104, 193, 194, 210
Bhaskaracharya, 171, 174
Bhatnagar, Shanti Swarup, 192, 193, 194, 196, 210, 214
Bhor Ghat, 106, 112, 205
Bhutan, 52
Bihar, 22, 41, 137, 166; earthquake in, 153, 206
Bihar Scientific Society, 157
biochemistry, 161
Birdwood, George, 125–6, 127

227

INDEX

Blanford, H. F., 134
Blyth, Edward, 30, 44
Board of Scientific Advice, 135–6
Board of Scientific and Industrial Research, 196
Bohr, Neils, 169, 174
Bombay, 7, 29, 33, 41, 62, 63, 64, 65, 74, 75, 88, 94, 106, 107, 108, 112, 113, 124, 125, 127, 142, 143, 144, 146, 158, 181, 205; dockyards in, 102–3; Government of, 75; University of, 191; textile mills in, 94, 125, 127
Bombay Educational Society, 33
Bombay Literary Society, 30, 32
Bombay Medical and Physical Society, 29, 77
Bombay Medical Service, 58, 125
Bose, Girindra Shekar, 176
Bose, Jagadis Chandra, 140, 153, 154, 155–6, 159, 163, 165, 166–8, 169, 174–5, 176, 184, 192, 193, 194, 214
Bose, Pramatha Nath, 139, 153–4, 155, 159, 160, 163, 171–2, 190, 214
Bose, Satyendra Nath, 192
Bose, Subhas Chandra, 207, 208
botanical gardens, 47–8, 49, 51–5
Botanical Survey of India, 49, 134, 138, 189, 197
botany, 10, 12, 13, 19, 20, 21, 22, 29, 38–9, 44, 46–50, 55, 60, 61, 131, 133–4, 137, 148–9, 151, 152, 160, 163, 175, 191–2
Brahmachari, Upendranath, 198
Brahmaputra, 104
Brahmins, 8, 36, 37, 55, 64, 72, 122, 154, 193
Brahmo Samaj, 124, 154, 173
Brandis, Dietrich, 39, 148–9, 214
British Association for the Advancement of Science, 25, 31, 162, 195
British Geological Survey, 46
British Trigonometrical Survey, 41
Brown, Alexander Crum, 155
Bryden, James L., 82, 83
Buchanan, Francis, 22, 23–4, 25, 28, 38–9, 44, 45, 48, 49, 77, 97, 214
Buckle, H. T., 80
Bundelkhand, 26, 99
Burdwan fever, 71, 79
Burkhill, I. H., 49
Burma, 41, 48
Butler, Fanny, 87
Butler, Sir Harcourt, 144

Calcutta, 7, 24, 27, 29, 30, 33, 36, 41, 45, 47–9, 59, 64, 73, 77, 78, 84, 90, 104, 106, 107, 109, 113, 114, 115, 124, 126, 134, 146, 153–4, 178; as a centre for science, 153–6, 158, 159–60, 162, 167; University of, 64, 191, 192
Calcutta Botanic Garden, 23, 47–8, 51–2, 134, 195
Calcutta International Exhibition, 181
Calcutta Journal of Natural History
Calcutta Madrassa, 62, 63
Calcutta Medical College, 63, 64, 114, 134, 142, 145, 153, 163, 181, 201
Calcutta Medical and Physical Society, 29, 31
Calcutta Review, 28, 32, 37
Calcutta School of Tropical Medicine, 145, 182, 186, 198
Calder, James, 44
Calicut, Zamorin of, 180
Cama hospital, 88
Cameron, W., 93
Campbell, Sir Colin, 114
Canning College, 162
Cannon, Susan Faye, 39
Caraka, 62, 173, 175, 176
Carey, William, 48–9
Carter, Henry Vandyke, 142
Cautley, Proby T., 21, 26–7, 117, 118, 119, 214
Cauvery, 116, 117, 205
Central Provinces, 107
Central Research Institute, 144, 188
Ceylon (Sri Lanka), 47, 51
Chandrasekhar, S., 193, 194
charka, see spinning
Charles, T. E., 73
Chatterjee, Partha, 17
Chelmsford, Lord, 186
chemistry; 5, 12, 35, 60, 151, 155, 160, 161, 162, 163–6, 191–3, 195; Hindu, 172–3
China, 4, 5, 10, 12, 14, 48, 53, 95
cholera, 57, 66, 68, 74, 78, 81–6, 123, 141, 142, 143, 144, 178, 198
Chopra, R. N., 182, 198, 214
Christianity, 48, 117, 124
Christophers, S. R., 145
Chuckerbutty, S. G., 64
Church of England Zenana Missionary Society, 87
cinchona, 48, 66, 134, 150
Cleghorn, Hugh, 148
Clow, A. G., 187
coal, 45–6, 98, 100, 107–8
Coal Committee, 45
Cochin, Maharaja of, 179
Coimbatore, 115, 205
Colebrooke, H. T., 37
colonial science, 7, 9–15, 27, 28–30, 39, 43, 50, 129, 147, 158, 211
Company science, 19, 19–56 *passim*, 129, 132, 147, 148, 211
Contagious Diseases Act, 90
Coonoor, 210, 202
Corbett, A. F., 120, 121
Cornish, W. R., 83
Cornwallis, 59

228

INDEX

cotton, 47, 48, 52, 116, 150, 150; mills, 93–4, 96–7, 125, 127
Cotton, Arthur, 116–17, 119, 120
Council of Scientific and Industrial Research, 196
Court of Directors, 19, 20, 21, 23, 25, 46, 47, 52, 58, 66, 102, 117
Cuningham, J. M., 59, 83
Cunningham, D. D., 83, 142, 155
Curzon, Lord, 15, 114, 119, 127, 135–7, 138, 144, 145, 151, 160, 161, 164, 186

Dacca, 7, 78, 115; Nawab of, 180; University of, 191, 192
Daji, Bhau, 70
Dalhousie, Lord, 109, 111, 114–15
Darjeeling, 55, 80, 112
Darwin, Charles, 19, 20, 27, 30, 170
Davy, Humphrey, 168
Dawn, 159, 170
Dawn Society, 158
Day, Francis, 133
Deakin, Arthur, 118, 119, 122
Deb, Radhakanta, 33
Dehra Dun, 41, 146, 148
Delhi, 6, 7, 8, 98, 112, 117, 153, 180–1, 199; King of, 124
Delhi Female Medical Mission, 87
Dempster, T. E., 81, 120
Department of Agriculture, 151
Desmond, Ray, 21
Dey, Kani Lal, 181–2
Dharampal, 99
diet, 79, 152, 201–2
Donnan, F. G., 193
d'Orta, Garcia, 46
Dufferin, Lady, 88
Dufferin Fund, 88–9, 90, 200
Dundas, Henry, 60
Durand, H. M., 26, 27
Dutt, R. C., 80
Dutt, U. C., 181

East India Company, 14, 40, 42–3, 44, 45–6, 48, 57, 58, 62, 85, 87, 94, 104, 106, 107; as a patron of science, 19–21, 24–5, 30
East India Railway, 112
Edinburgh University, 49, 60, 140, 154, 155, 194
Edney, Matthew, 31, 40, 42–3
Egypt, 118
Einstein, Albert, 169, 174, 192
electricity, 97, 121, 124, 166
Elgin, Lord, 134, 142
Elphinstone College, 33
engineering, 15, 192, 194
entomology, 144, 195, 199

Epidemic Diseases Act, 143
ethnography, 24, 97, 130
ethnology, 36, 131
Everest, George, 27, 28, 41, 214

Falconer, Hugh, 21, 26–7, 28, 49, 50, 214
famine, 47, 51, 83, 86, 93, 108, 121, 123, 127, 134, 135, 147, 151, 202–3; and irrigation 116, 117, 119, 120–1, 134
Far Eastern Association of Tropical Medicine, 198
Farr, William, 84
Fayrer, Sir Joseph, 82
fisheries, 23, 133, 197
forestry, 137, 147–50, 187, 197
Foucault, Michel, 23, 40
Franklin, James, 26, 28

Gadgil, Madhav, 147
Gandhi, M. K., 17, 126, 127–8, 189–90, 206–7, 208
Ganges, 104, 105, 112, 174
Ganges Canal, 117–19, 123
Geddes, Patrick, 166
Geological Society of London, 19, 29, 30
Geological Survey of India, 25, 26–7, 30, 133, 137, 138, 139, 164, 193
Geological Survey of Ireland, 25, 46
geology, 10, 13, 19, 26–8, 29, 36, 44–6, 55, 61, 131, 137, 158, 160, 163, 191–2
Ghosh, J. C., 192, 193, 209, 210
Ghosh, Jogendranath, 159
Ghosh, P. C., 126
Ghosh, Rash Bihari, 160
Gibson, Alexander, 148
Goa, 6, 46
Godavari, 116, 117
Goldingham, John, 36, 41
Gopalacharlu, D., 178
Gosling, David, 170
Govan, George, 48
Government of India, 25, 30, 59, 60, 75, 83, 120, 130, 188, 196, 198, 204; attitudes to medicine and public health, 142–4; attitudes to science, 132, 133–4, 135–7, 139, 156
Government of India Act, 1919, 183, 186, 188
Government of India Act, 1935, 208
Government School of Art, 126
governmentality, 23, 130
Grand Trunk Road, 106–7
Grant, Charles, 125
Grant, John B., 195
Grant Medical College, 63, 64, 69, 70, 146
Great Trigonometrical Survey of India, 25, 39, 41–3, 46, 114, 119, 138
Greece, 3, 56

229

INDEX

Greeks, 3, 173, 184
Grierson, D., 156
Grout, Andrew, 44
Grove, Richard, 147, 148
Guha, Ramachandra, 147
Gujarat, 22, 94, 97, 101, 115

Haffkine, Waldemar, 142, 143, 145
Haffkine Institute, 144, 195
hakims, 4, 67–8, 69, 71, 180–1, 183, 184
Halbfass, Wilhelm, 170
Hankin, E. H., 142, 145
Hardinge, Lord, 109
Hardinge College, 89
Hardwar, 83, 118
Hare School, 153
Hart, Ernest, 141
Havell, E. B., 125, 126–7
Health Survey and Development Committee (Bhore Committee), 204
helminthology, 144
Herbert, J. D., 30, 36
Heyne, Benjamin, 67
Hill, A. V., 196–8, 203
hill-stations, 55, 146–7, 153, 188, 199
Himalayas, 39, 41, 45, 48, 50, 52, 53, 55, 78, 98, 117, 118
Hindu College, 153
Hindu medicine, 7, 55, 67, 177, 180; *see also* Ayurveda
Hindu science, 2–4, 36–8, 169–76
Hodgson, B. H., 24, 44, 49, 55, 215
Holland, Sir Thomas, 139, 155, 164, 215
Holland Commission, *see* Indian Industrial Commission
Holwell, J. Z., 72
homoeopathy, 176, 178
Hooghly, 108
Hooker, Joseph Dalton, 20, 21, 27, 39, 53, 131, 175, 215
hookworm, 145, 194
horticulture, 5, 47
Hove, Anton, 22, 97
Howard, Albert, 138, 152, 153, 215
Howard, Gabrielle, 152
Humboldt, Alexander von, 20, 28, 44, 52, 53
Hunter, W. W., 122, 130–1
Hyderabad, 6, 160
hydro-electricity, 8, 205

Imperial Bacteriological Laboratory, 142, 146
Imperial Council of Agricultural Research, 186, 187
Imperial Gazetteer, 130–2
Imperial Institute, 161

imperial science, 15, 114, 135–9, 144
improvement, doctrine of, 23, 33, 45, 47, 56, 119, 125
Inchcape, Lord, 187
Indian Academy of Sciences, 195
Indian Advisory Committee, 135–6
Indian Agricultural Research Institute, 137, 151–3, 163
Indian Agricultural Service, 138, 151, 186
Indian Association for the Cultivation of Science, 153, 158–9
Indian Botanical Society, 194
Indian Chemical Society, 194–5
Indian Civil Service, 58, 59, 129–33, 137, 154, 156, 164, 183, 187, 188, 190
Indian Educational Service, 138, 140, 160, 165, 166
Indian Famine Code, 134
Indian Famine Commissions, 108, 134, 151
Indian Forest Acts, 147
Indian Forest Service, 138, 147, 148, 150
Indian Industrial Association, 159
Indian Industrial Commission, 138–9, 150, 164, 186
Indian Institute of Science, 8, 161, 195
Indian Irrigation Commission, 117, 119, 120
Indian Journal of Medical Research, 144
Indian Medical Congress, 141, 142, 181–2
Indian Medical Service, 57–61, 64–5, 89, 119, 137, 141–2, 144–5, 156, 160, 163, 175, 183, 185, 188, 198–9, 210, 213
Indian National Congress, 126, 183, 184, 189, 207, 208
Indian Plague Commission, 143
Indian Red Cross, 200
Indian Research Fund Association, 144, 188–9, 198, 200, 201, 205, 210
Indian Retrenchment Committee, 187
Indian School of Forestry, 146, 148, 150
Indian Science Congress, 29, 161–2, 166, 169, 173, 189, 195, 210
Indian Universities Act, 160
Indianisation, 60–1, 64–5, 138, 139–40, 195
Indigenous Drugs Committee, 182
indigenous medicine, 62–3, 65–71, 77; revival of, 176–85; *see also* Ayurveda, Unani medicine
Indus, 104
Industrial Health Advisory Committee, 201
Industrial Research Fund, 196
Iran, 5, 44
Irish Ordnance Survey, 41
irrigation, 23, 115–21, 134
irrigation canals, 45, 48, 108, 150; construction of, 116–20, 122, 123; environmental effects of, 81, 120–1

230

INDEX

Jacquemont, Victor, 20
Jai Singh, Raja, 6, 36
Jaipur, 6
Jambedkar, Bal Shastri, 32
Jameson, James, 82
Jamshedpur, 139
Japan, 10, 11, 155, 166
Jayakar, M. R., 188
Jeejeebhoy, Jamsetji, 64
Jenner, Edward, 72
Jerdon, T. C., 44
Jessore, 80
Jesuits, 6, 154
Johnson, James, 77
Jones, Sir William, 29, 30, 32, 34–5, 38, 55, 56, 66, 97, 161, 171
Julahas, 94
Jumna Canal, 26, 117, 120

kala-azar, 144, 198
Kala Bhavan Technical Institute, 160
Kanara, 22
Kanpur, 106
Karachi, 41
Karnal, 106
Kasauli, 142, 144, 145, 146, 198–9
Kashmir, 50
Kashyap, Shiv Ram, 194
Kayasthas, 8, 154
Kaye, Sir John, 122
Kelvin, Lord, 167
Kerala, 179–80
Kerr, Ian, 112, 113
Kew Gardens, 22, 134
Khan, Hakim Ajmal, 180–1
Khan, Sir Sayyid Ahmad, 157
Khan, Sharif, 68
Kidderpore, 102
King, George, 134, 155
King Institute of Preventive Medicine, 144
Kirtikar, K. R., 175
Kittredge, George A., 88
Klein, Ira, 86
Koch, Robert, 57, 84, 141, 142
Kodaikanal, 134
Koenig, J. G., 46
Kolar, 99, 205
Kopf, David, 62
Krishnagar College, 154
Krishnan, K. S., 193, 194
Kumaon, 146
Kumar, Deepak, 139, 148
Kurumbars, 150
Kyd, Robert, 47–8, 49, 51, 150

Lafont, Father, 154
Lahore, 7, 139, 157, 175, 192; University of, 191, 194
Lal, Maneesha, 89
Lambton, William, 41, 42, 215
Lansdowne, Lord, 146
Larwood, H. J. C., 25, 45
Laveran, Alphonse, 141
League of Nations, 194, 198, 202
Legislative Assembly, 188
leprosy, 70, 142
Leslie, Charles, 177
Lewis, T. R., 83, 142
Linlithgow, Lord, 201
Linnaeus, 19, 38, 44, 46
Lister, Lord, 167
Liston, Glen, 145
London Missionary Society, 87
Lucknow, 6, 7, 36, 37, 114, 180, 184; University of, 191, 192
Lukis, Sir Pardy, 144, 145, 183
Lyell, Charles, 19, 20, 27, 28, 46

Macaulay, T. B., 38, 40
McCarrison, Robert, 201, 202, 215
McCay, D., 201
MacGeorge, G. W., 108, 109, 122, 203
Mackenzie, Colin, 20, 23, 25, 40, 42, 215
M'Clelland, John, 31, 45–6
MacLeod, Roy, 136
MacMahon, P. S., 162
M'Cosh, John, 77
Madras, 7, 19, 29, 41, 65, 67, 113, 144, 178, 182; famine in, 108; Government of, 42, 184; Observatory in, 36, 41, 134; province of, 23, 65, 74, 77, 126, 158; University of, 154, 191, 192
Madras Medical College, 62, 88, 185
Madras Medical Service, 58
Madrassa Tibbia, 180
Mahalanobis, P. C., 194, 210
Malabar, 45, 46, 51, 101, 150, 179
malaria, 57, 66, 68, 69, 71, 78–81, 86, 91, 120, 130, 141–2, 144, 145, 178, 189, 198–9, 201
Malaria Institute, 199, 205
Malkowal, 142
Manson, Patrick, 141, 142
mapping, 36, 40–3
Marathas, 6, 19, 22, 41, 42, 101, 106
Martin, James Ranald, 63, 69, 77, 85, 86, 215
Marx, Karl, 111
Masulipatam, 19, 101
materia medica, 4, 50, 66, 67, 70, 175, 181–2
Mathematical Society, 194
mathematics, 3, 12, 33, 160, 162, 171, 191–2

Mathura, 6
Mayo, Lord, 151
Mayurbhanj, Maharaja of, 139
Medical Research Department, 188
medical topography, 69, 75–9, 81, 82
medicine, 3, 5, 15, 22, 23, 35, 57–91 *passim*, 127, 136–7, 141–6, 147, 151, 163, 192, 204; connections with botany, 49–50, 66; *see also* allopathy, indigenous medicine
Medlicott, H. B., 139
Meerut, 106, 118
Megaw, J. W. D., 203
metallurgy, 5, 100–66
metals, 97–9
Meteorological Department, 134, 137, 138
meteorology, 28, 36, 82, 83, 131, 134, 171
Mewar, 101
mineralogy, 44, 46
mines and mining, 23, 98–101
missionaries, 16, 87–8, 89, 126, 156
Mitra, Rajendralal, 158
modernity, 8, 15–18, 128, 169, 206, 209–10
Montagu, Edwin, 186
Morehead, Charles, 69, 70, 82
Morley, Lord, 65
Mughals, 3, 5, 6, 8, 106, 117, 180
Mukherjee, Asutosh, 160, 161, 162
Mukherjee, J. N., 193
Mukherjee, T. N., 181
Mukherji, D. N., 152
Mukherji, U. N., 80
Muktesar, 137, 146, 186
museums, 29–30, 32
Muslims and Indian science, 3, 4–6, 173, 177
Mutiny/Rebellion, 1857, 6, 83, 84, 101, 114, 121, 122–4, 132, 154
Mysore, 6, 8, 22–3, 40, 42, 99, 160, 161, 205

Nadia, 6
Nandy, Shib Chandra, 115
Narmada, 107–8
National Council of Education, 160
National Institute of Sciences of India, 195–6, 197, 209
National Planning Committee, 208, 209
national science, 10–11, 14, 158–60, 164–8, 207–10
Native Medical Institutions, 62–3, 185
Needham, Joseph, 5
Nehru, Jawaharlal, 15, 17, 18, 206, 207, 208, 209–10
Nehru, Motilal, 206
Nepal, 48
Newbold, T. J., 44, 45
Nightingale, Florence, 84, 85, 86
Nilgiris, 53, 78

Nobel prizes, 8, 144, 167, 168, 169, 212
North-Western Provinces, 78, 108, 120–1, 158
numismatics, 28, 36, 155
nutrition, 201–2, 203

O'Donnell, C. J., 80
Oldham, R. D., 155
Oldham, Thomas, 25, 28, 46
O'Malley, L. S. S., 80
Ootacamund, 112, 113
Orientalism, 3, 4, 34–9, 50, 55–6, 62–3, 66, 67, 170, 177
Orissa, 83, 134, 151
ornithology, 50
O'Shaughnessy, W. B., 68, 114–15, 163, 215
Osmania University, 191

Palit, Tarakanath, 160
Panagal, Raja of, 184
Pandarpur, 83
Panikkar, K. N., 180
Parsis, 8, 64, 88, 102, 103–4, 161, 193
Pasteur, Louis, 141, 142, 209
Pasteur Institutes, 142, 144
Pechey, Edith, 88
Peddler, Alexander, 140
Peel, Sir Robert, 109
Peshawar, 106
pharmacology, 4, 50, 69, 70, 71, 180, 181–2, 198
philanthropy, 160–1, 163
philology, 36, 155
Phipps, Henry, 151
physics, 12, 155, 158, 160, 161, 162, 166–8, 169, 191–3
physiology, 185, 191, 195, 201, 202; plant, 167–8, 174, 194
pilgrimages, 83, 123, 143
plague, 57, 75, 86, 123, 135, 141, 142–4, 145, 178, 198
Playfair, George, 68
Playfair, John, 68
Pogson, N. R., 134
Poona, 36, 112, 142, 146
Prain, D., 154, 155
Prakash, Gyan, 16, 17
Prasad, Vishnu, 49
Presidency College, Calcutta, 140, 154, 165, 166, 175
Presidency College, Madras, 162
Prinsep, James, 28, 36, 130
psychology, 177, 192, 195
Punjab, 6, 42, 69, 74, 90, 106, 107, 109, 119, 120, 143, 144, 158, 206; University of, 191
Puri, 83
Pusa, 137, 138, 151–3, 163, 186

INDEX

Pye, Lucien, 16
Pykara, 205

Quetta, 152, 206

race, 79–81, 131, 139–40
racial discrimination, 59, 139–40, 145–6, 155
railways, 15, 45, 46, 121, 122, 134, 150, 205; construction of, 102–3, 108–9, 111–13, 117, 119; social and economic impact of, 96, 100, 105–6, 107, 109–11, 113–14, 119, 127
Rajasthan, 98, 101, 115
Ramakrishna Mission, 200
Raman, Chandrasekhara Venkata, 8, 154, 155, 159, 161, 162, 167–8, 169, 189, 190, 192, 193, 194, 195, 212, 215
Ramanujan, Srinivasa, 155, 162, 194
Rampur, Nawab of, 180
Rangoon, 114
Raniganj, 100, 108
Rankine, Robert, 77
Rawalpindi, 111
Ray, Prafulla Chandra, 140, 153, 154, 155, 156, 159, 162, 163, 164–5, 172–3, 189, 190, 192, 193, 194, 215
Reed, Sir Stanley, 203
Rennell, James, 40
Risley, H. H., 131
road-building, 106–7
Rockefeller Foundation, 194
Rogers, Sir Leonard, 145
Rome, 3, 56, 107, 119, 122
Roorkee, 118
Ross, Ronald, 57, 80, 132, 141, 142, 145, 198, 216
Roy, Gopaul Chunder, 80
Roy, William, 42
Royal Army Medical Corps, 59, 142
Royal Asiatic Society, 29, 70
Royal Commission on Agriculture in India, 110, 186, 201, 203
Royal Commission on the Sanitary State of the Army in India, 84–5
Royal Engineers, 116, 119, 138
Royal Institution, 166, 168
Royal Navy, 102
Royal Society, 19, 29, 36, 52, 135–6, 167, 193, 195, 196; Fellows of, 27, 103, 193, 196
Roxburgh, William, 20, 24, 44, 47, 48, 49, 66, 175, 216
Royle, J. Forbes, 28, 48, 49, 50, 52–3, 55, 67, 150, 175, 216
Ruskin, John, 125, 127

Saha, Meghnad, 190, 192, 193, 194, 195, 207, 208, 209, 216
Saharanpur Botanic Garden, 26, 48, 49–50, 52–5

Sahni, Birbal, 168, 193, 194
St Xavier's College, 154
Salem, 96, 100, 115
Sangwan, Satpal, 103
Sanskrit, 6, 35, 38, 39, 55, 66, 154, 174, 177, 178, 185
Sanskrit College, 62, 63
Saraswati, Dayananda, 170
Sarkar, Nilratan, 160
Science and Culture, 207, 209
scientific communities, 10, 14, 153–63, 170
scientific journals, 30–1, 157, 165
scientific Romanticism, 168, 174
Seal, Brajendranath, 173–4, 176, 184
Sen, Chandra Kishore, 178
Sen, Gananth, 177, 184
Sen, Jatindra Nath, 153
Sen, Keshab Chandra, 124, 170
Sen, Ramkamal, 32
Serampore, 48
Serfoji, 6
Seringapatnam, 6
Shah Jehan, 117
Sheriff, Mohideen, 181
shipbuilding, 5, 101–3
Sholapur, 96
Siddha, 176, 185
Sikdar, Radhanath, 43
Sikhs, 101, 106, 139
Simms, E. W., 111
Simon Commission, 188
Simonsen, J. L., 162
Sinclair, Sir John, 23
Sind, 105, 109
Singh, Laksman, 49
Singh, Ranjit, 5
Sinhji, Bhagvat, 177
Sircar, Mahendralal, 153–4, 155, 158–9, 163, 176, 189, 216
Sitala, 73–4, 76
Sivasamudram, 205
Siwalik fossils, 21, 26–7, 28, 45
smallpox, 71–5, 76, 198
Snow, John, 82
Society for the Promotion of Technical Education, 160
Society for the Propagation of the Gospel, 87
Society for Translating European Sciences, 158
spinning, 94, 95, 128, 190
Srilal, Pandit, 125
Srinivasa Murti, G., 184–5
steamships, 15, 30, 45, 96, 102–3, 104–5, 106, 114, 124, 127
Stone, Ian, 120–1
Strong, F. P., 78
Suez Canal, 96, 114

INDEX

Surat, 19, 101, 102
surgeons (East India Company), 25, 26, 27, 45, 49, 57, 60, 61, 147
surveys, 22–4, 39–43, 123, 130
Surya Siddhanta, 3, 37
Susruta, 62, 173
Swadeshi movement, 127, 159–60, 165, 172
Swain, Clara, 87
Sykes, W. H., 130

Tagore, Dwarkanath, 33, 100
Tagore, Rabindranath, 169
Tait, P. G., 155
Tamilnadu, 115
Tanjore, 6, 116
Tantimul, Lalla, 106
Tata, J. D., 104
Tata, Jamsetjee Nusserwanjee, 161
Tata iron and steel works, 139
taxonomy, 38–9, 46
Taylor, James, 78
telegraphs, 113–15, 122–3, 163
textiles, 5, 22, 93–7, 103, 126–7
Thornton, William, 107
Tibet, 43
Tinnevelly, 115
Tipu Sultan, 6, 42, 106
Tod, James, 100
Topping, Michael, 36, 41
Trautmann, Thomas, 35, 37
Travancore, 160; Maharaja of, 179
trigonometrical surveys, 13, 39–43, 44
Trivandrum, 36
tropical disease and medicine, 61, 68–9, 141, 144, 145, 198–9
tropics, ideas of, 50–3, 55–6, 77, 146, 178
Twining, William, 68, 79

Udaipur, 26
Ujjain, 6
Unani medicine, 6, 7, 62–3, 66, 67, 69, 71, 176, 177, 178, 180–1, 183, 185, 204
United Provinces, 120
United States, 10, 11, 118, 155, 196, 213

universities, science at, 160, 191–4, 197, 205
University College London, 193
University College of Science 160, 163, 165, 191
Urdu, 62, 157
Usman, Mohammad, 184, 185

vaccination, 66, 71–5, 123
vaids, 4, 67–8, 178, 179, 180, 183
Vaidyas, 8, 63, 154
Van Rheede, Heinrich, 46
Varier, P. S., 179–80
variolation, 71–4
Vernacular Literary Society, 158
veterinary science, 137, 138, 187
Victoria, Queen, 88, 89, 121–2
Vicziany, Marika, 22, 23
Vivekanada, Swami, 170
Vizagapatam College, 154
Vizianagram, Raja of, 156
Voelcker, J. A., 133, 151, 152

Wadia, Ardaseer Cursetjee, 103, 193
Wadia, D. N., 193
Wadia, Lowji Nuserwanji, 102
Wallich, Nathaniel, 29, 48, 49, 216
Waring, E. J., 181
Watson, E. R., 192, 195
Watt, George, 132, 181, 182
Wellesley, Lord, 22, 106
West Indies, 51, 52, 77, 138
Whitcombe, Elizabeth, 121
White, Lynn, 5
Wight, Robert, 39
Williams, D. H., 46
Wilson, H. H., 67
Wise, T. A., 67, 176
Women's Christian Medical College, 88
Women's Medical Service, 89, 199
Wudders, 112

Zenana Bible and Medical Mission, 87
Zoological Survey of India, 138, 189, 197
zoology, 10, 12, 19, 24, 29, 30, 36, 44, 49, 55, 131, 160, 162, 175, 191–2

THE NEW CAMBRIDGE HISTORY OF INDIA

I The Mughals and Their Contemporaries

*M.N. Pearson, *The Portuguese in India*
*Burton Stein, *Vijayanagara*
*Milo Cleveland Beach, *Mughal and Rajput Painting*
*Catherine Asher, *Architecture of Mughal India*
†*John F. Richards, *The Mughal Empire*
*George Michell, *Architecture and Art of Southern India*
*George Michell and Mark Zebrowski, *Architecture and Art of the Deccan Sultanates*
Richard M. Eaton, *Social History of the Deccan*
Bruce R. Lawrence, *Indian Sufism and the Islamic World*

II Indian States and the Transition to Colonialism

†*C.A. Bayly, *Indian Society and the Making of the British Empire*
*P.J. Marshall, *Bengal: the British Bridgehead: Eastern India 1740–1828*
†*J.S. Grewal, *The Sikhs of the Punjab*
*Stewart Gordon, *The Marathas 1600–1818*
*Om Prakash, *European Commercial Enterprise in Pre-Colonial India*
Richard B. Barnett, *Muslim Successor States*
David Washbrook, *South India*

III The Indian Empire and the Beginnings of Modern Society

*Kenneth W. Jones, *Social and Religious Reform Movements in British India*
*Sugata Bose, *Peasant Labour and Colonial Capital: Rural Bengal since 1770*
†*B.R. Tomlinson, *The Economy of Modern India, 1860–1970*
†*Thomas R. Metcalf, *Ideologies of the Raj*
*David Arnold, *Science, Technology and Medicine in Colonial India*
Gordon Johnson, *Government and Politics in India*
B.N. Ramusack, *The Indian Princes and Their States*

IV The Evolution of Contemporary South Asia

†*Paul R. Brass, *The Politics of India since Independence: Second Edition*
*Geraldine Forbes, *Women in Modern India*
*Susan Bayly, *Caste, Society and Politics in India from the Eighteenth Century to the Modern Age*
*David Ludden, *An Agrarian History of South Asia*
Rajnarayan Chandavarkar, *The Urban Working Classes*
Francis Robinson, *Islam in South Asia*
Anil Seal, *The Transfer of Power and the Partition of India*

* *Already published*   † *Available in paperback*

LANCASTER UNIVERSITY

LIBRARY

Due for return **b**y:
end of service on date below (or
earlier if recalled)

-7 FEB 2003

2 2 NOV 2004

25 JAN 2005